Workbook

SRA

# Subtraction

A Direct Instruction Program

Siegfried Engelmann • Doug Carnine

Columbus, OH

The McGraw·Hill Companies

**Cover and title page photo credits:**
©Steve Mercer/Getty Images, Inc.

SRAonline.com

Send all inquiries to:
SRA/McGraw-Hill
8787 Orion Place
Columbus, OH 43240-4027

Printed in the United States of America.

ISBN 0-07-602462-8

3 4 5 6 7 8 9 VHG 08 07 06

The McGraw·Hill Companies

# Subtraction Preskill Test

## A

| | | | | | | |
|---|---|---|---|---|---|---|
| 8<br>+ 9 | 7<br>+ 5 | 8<br>+ 4 | 7<br>+ 9 | 6<br>+ 8 | 5<br>+ 9 | 8<br>+ 3 |
| 7<br>+ 6 | 6<br>+ 6 | 8<br>+ 7 | 9<br>+ 2 | 8<br>+ 8 | 5<br>+ 8 | 7<br>+ 4 |
| 8<br>+ 5 | 9<br>+ 4 | 7<br>+ 7 | 9<br>+ 7 | 6<br>+ 5 | 9<br>+ 9 | 8<br>+ 4 |

## B

| | | | | |
|---|---|---|---|---|
| 351<br>+ 120 | 47<br>+ 25 | 605<br>+ 123 | 86<br>+ 13 | 264<br>+ 183 |

## C

Ann buys 5 tires. She buys 2 more tires. How many tires in all did Ann buy?

George found 6 cans. He found 3 more cans. How many cans in all did George find?

Chris got 4 pens. She got 1 more pen. How many pens in all did Chris get?

Jose made 7 chairs. He made 2 more chairs. How many chairs in all did Jose make?

# Subtraction Preskill Test

## A

| | | | | | | |
|---|---|---|---|---|---|---|
| 8 + 9 | 7 + 5 | 8 + 4 | 7 + 9 | 6 + 8 | 5 + 9 | 8 + 3 |
| 7 + 6 | 6 + 6 | 8 + 7 | 9 + 2 | 8 + 8 | 5 + 8 | 7 + 4 |
| 8 + 5 | 9 + 4 | 7 + 7 | 9 + 7 | 6 + 5 | 9 + 9 | 8 + 4 |

## B

| | | | | |
|---|---|---|---|---|
| 351 + 120 | 47 + 25 | 605 + 123 | 86 + 13 | 264 + 183 |

## C

Ann buys 5 tires. She buys 2 more tires. How many tires in all did Ann buy?

George found 6 cans. He found 3 more cans. How many cans in all did George find?

Chris got 4 pens. She got 1 more pen. How many pens in all did Chris get?

Jose made 7 chairs. He made 2 more chairs. How many chairs in all did Jose make?

# Subtraction Placement Test

## A

| | | | | |
|---|---|---|---|---|
| $7 - 1$ | $8 - 4$ | $5 - 1$ | $6 - 0$ | $6 - 3$ |
| $10 - 5$ | $4 - 0$ | $9 - 1$ | $4 - 2$ | $6 - 1$ |

## B

| | | | | | |
|---|---|---|---|---|---|
| $82 - 35$ | $54 - 37$ | $526 - 165$ | $426 - 318$ | $62 - 5$ | $347 - 62$ |

## C

There are 148 red cars and 432 blue cars. How many more blue cars are there than red cars?

The shop gave away 86 red balloons. The shop gave away 90 blue balloons. How many balloons in all did the shop give away?

Part C continues on the next page.

Lucy found 164 pencils. 98 of the pencils were broken. How many of the pencils were not broken?

1840 girls go to our school. There are 3000 children altogether in our school. How many boys go to our school?

When Bill started high school, he weighed 108 kilograms. He has gained 24 kilograms since then. How many kilograms does Bill weigh now?

At the beginning of the week, Jenny had 945 points. At the end of the week, Jenny had 1000 points. How many points did Jenny get during the week?

The truck is carrying 1463 kilograms of apples and 3652 kilograms of oranges. How many kilograms of fruit is the truck carrying?

The old car weighs 986 kilograms. The new car weighs 2000 kilograms. How much heavier is the new car?

# Lesson 1

Bonus

## 1

| | | | |

## 2

| A | B | C | D | E | F |
|---|---|---|---|---|---|
| 264 | 538 | 492 | 751 | 892 | 376 |

## 3

**A**

7 { 4 — $7 - 4 = 3$

3 — $7 - 3 = 4$

**B**

9 { 2 ________

7 ________

**C**

3 { 2 ________

1 ________

**D**

5 { 3 ________

2 ________

# Lesson 2

Bonus

**1**

**A** 279 **B** 146 **C** 657 **D** 328 **E** 864

**2**

| | | | | |

**3**

| 4 | 6 | 9 | 5 | 8 | 10 | 7 | 2 |
|---|---|---|---|---|---|---|---|
| − 1 | − 1 | − 1 | − 1 | − 1 | − 1 | − 1 | − 1 |

**4**

**A** 8 { 5 ____ 8 − 5 = 3 ____; 3 ____ 8 − 3 = 5 ____

**B** 6 { 4 ______________; 2 ______________

**C** 5 { 3 ______________; 2 ______________

**D** 7 { 2 ______________; 5 ______________

**5**

**A** ____ 5 − 5 = 0 ____

**B** ______________

**C** ______________

**D** ______________

**E** ______________

# Lesson 3

Bonus

## 1

| A | B | C | D | E |
|---|---|---|---|---|
| 425 | 738 | 916 | 437 | 514 |

## 2

$5 - 1 =$ ____  $7 - 1 =$ ____  $10 - 1 =$ ____  $8 - 1 =$ ____  $4 - 1 =$ ____  $9 - 1 =$ ____  $6 - 1 =$ ____  $2 - 1 =$ ____

## 3

**A** [ ] { ___ + ___ = ___ ; 1 ___ − ___ = ___ ; 4 ___ − ___ = ___ }

**B** [ ] { ___ + ___ = ___ ; 1 ___ − ___ = ___ ; 6 ___ − ___ = ___ }

**C** [ ] { ___ + ___ = ___ ; 1 ___ − ___ = ___ ; 8 ___ − ___ = ___ }

**D** [ ] { ___ + ___ = ___ ; 1 ___ − ___ = ___ ; 3 ___ − ___ = ___ }

## 4

**A** $6 - 6 = 0$

**B** ________________

**C** ________________

**D** ________________

**E** ________________

**F** ________________

# Lesson 4

Bonus

## 1

A 356 B 297 C 424 D 932 E 358 F 672

## 2

A 10: 5, 5 — 5 + 5 = 10; 10 − 5 = 5

B 18: 9, 9 — +; −

C 6: 3, 3 — +; −

D 8: 4, 4 — +; −

## 3

$6 - 1$ $8 - 1$ $2 - 1$ $9 - 1$ $4 - 1$ $7 - 1$ $10 - 1$ $3 - 1$

## 4

A [ ]: 1, 4 — +; −; −

B [ ]: 1, 7 — +; −; −

C [ ]: 1, 9 — +; −; −

D [ ]: 1, 3 — +; −; −

# Lesson 5

| Facts | + | Bonus | = | TOTAL |
| --- | --- | --- | --- | --- |
| | | | | |

## 1

| A | B | C | D | E | F |
| --- | --- | --- | --- | --- | --- |
| 406 | 308 | 728 | 507 | 346 | 205 |

## 2

**A** 4 { 2, 2
- 2 + 2 = 4
- 4 − 2 = 2

**B** 10 { 5, 5
- +
- −

**C** 18 { 9, 9
- +
- −

**D** 6 { 3, 3
- +
- −

## 3

$10 - 5$   $6 - 3$   $6 - 1$   $18 - 9$   $6 - 3$   $8 - 1$   $7 - 1$   $18 - 9$

$2 - 1$   $10 - 5$   $6 - 3$   $8 - 1$   $18 - 9$   $10 - 5$   $10 - 1$   $7 - 1$

## 4

**A** ☐ { 1, 6
- +
- −
- −

**B** ☐ { 1, 9
- +
- −
- −

**C** ☐ { 1, 8
- +
- −
- −

**D** ☐ { 1, 3
- +
- −
- −

# Lesson 6

Facts + Bonus = TOTAL

## 1

| A | B | C | D | E | F |
|---|---|---|---|---|---|
| 508 | 326 | 409 | 583 | 704 | 542 |

## 2

A
14 { 7 ______ + ______
7 ______ − ______

B
6 { 3 ______ + ______
3 ______ − ______

C
18 { 9 ______ + ______
9 ______ − ______

D
10 { 5 ______ + ______
5 ______ − ______

## 3

$18 - 9$  $6 - 1$  $10 - 5$  $8 - 1$  $6 - 3$  $9 - 1$  $18 - 9$  $6 - 3$

$10 - 5$  $4 - 1$  $18 - 9$  $6 - 3$  $10 - 1$  $10 - 5$  $3 - 1$  $2 - 1$

## 4

A
[ ] { ______ + ______
1 ______ − ______
4 ______ − ______

B
[ ] { ______ + ______
1 ______ − ______
8 ______ − ______

C
[ ] { ______ + ______
1 ______ − ______
5 ______ − ______

D
[ ] { ______ + ______
1 ______ − ______
3 ______ − ______

# Lesson 7

Facts + Bonus = TOTAL

## 1

A 402

B 384

C 108

D 530

E 714

## 2

$2 - 0$   $5 - 0$   $7 - 1$   $4 - 0$   $8 - 1$   $6 - 0$   $8 - 0$   $4 - 1$

$3 - 0$   $8 - 1$   $6 - 0$   $4 - 0$   $10 - 1$   $3 - 1$   $6 - 0$   $2 - 0$

## 3

A 18 { 9, 9
+ ______
− ______

B 14 { 7, 7
+ ______
− ______

C 8 { 4, 4
+ ______
− ______

D 12 { 6, 6
+ ______
− ______

## 4

$6 - 3$   $18 - 9$   $4 - 1$   $7 - 1$   $10 - 1$   $10 - 5$   $18 - 9$   $6 - 1$

$6 - 3$   $9 - 1$   $10 - 5$   $18 - 9$   $6 - 3$   $10 - 1$   $10 - 5$   $9 - 1$

## 5

**A**

1 ____________________

8 ____________________

**B**

1 ____________________

6 ____________________

**C**

1 ____________________

9 ____________________

**D**

1 ____________________

3 ____________________

**E**

1 ____________________

5 ____________________

**F**

1 ____________________

7 ____________________

# Lesson 8

Facts + Bonus = TOTAL

## 1

$$\begin{array}{r}15\\-10\\\hline\end{array}\quad\begin{array}{r}13\\-10\\\hline\end{array}\quad\begin{array}{r}14\\-10\\\hline\end{array}\quad\begin{array}{r}12\\-10\\\hline\end{array}\quad\begin{array}{r}15\\-10\\\hline\end{array}\quad\begin{array}{r}11\\-10\\\hline\end{array}\quad\begin{array}{r}13\\-10\\\hline\end{array}\quad\begin{array}{r}12\\-10\\\hline\end{array}$$

## 2

**A** 3456

**B** 4546

**C** 2538

**D** 1324

**E** 8260

## 3

$$\begin{array}{r}6\\-0\\\hline\end{array}\quad\begin{array}{r}4\\-0\\\hline\end{array}\quad\begin{array}{r}3\\-1\\\hline\end{array}\quad\begin{array}{r}8\\-0\\\hline\end{array}\quad\begin{array}{r}10\\-1\\\hline\end{array}\quad\begin{array}{r}6\\-0\\\hline\end{array}\quad\begin{array}{r}4\\-1\\\hline\end{array}\quad\begin{array}{r}9\\-0\\\hline\end{array}$$

## 4

**A** 8 { 4 ____ + ____; 4 ____ − ____ }

**B** 18 { 9 ____ + ____; 9 ____ − ____ }

**C** 14 { 7 ____ + ____; 7 ____ − ____ }

**D** 10 { 5 ____ + ____; 5 ____ − ____ }

## 5

$$\begin{array}{r}14\\-7\\\hline\end{array}\quad\begin{array}{r}8\\-4\\\hline\end{array}\quad\begin{array}{r}18\\-9\\\hline\end{array}\quad\begin{array}{r}10\\-5\\\hline\end{array}\quad\begin{array}{r}6\\-3\\\hline\end{array}\quad\begin{array}{r}14\\-7\\\hline\end{array}\quad\begin{array}{r}18\\-9\\\hline\end{array}\quad\begin{array}{r}8\\-4\\\hline\end{array}$$

## 6

| | | | | | | | |
|---|---|---|---|---|---|---|---|
| 14 − 7 | 8 − 4 | 6 − 3 | 10 − 5 | 18 − 9 | 8 − 1 | 6 − 1 | 14 − 7 |
| 3 − 1 | 18 − 9 | 8 − 1 | 8 − 4 | 6 − 3 | 4 − 1 | 14 − 7 | 6 − 1 |
| 10 − 5 | 8 − 4 | 14 − 7 | 6 − 3 | 18 − 9 | 7 − 1 | 10 − 1 | 10 − 5 |
| 8 − 1 | 7 − 1 | 10 − 5 | 6 − 1 | 14 − 7 | 8 − 4 | 6 − 3 | 8 − 1 |

## 7

**A** ☐ { 0 ____ − ____ ; 8 ____ − ____ }

**B** ☐ { 0 ____ − ____ ; 6 ____ − ____ }

**C** ☐ { 1 ____ − ____ ; 3 ____ − ____ }

**D** ☐ { 0 ____ − ____ ; 9 ____ − ____ }

**E** ☐ { 1 ____ − ____ ; 5 ____ − ____ }

**F** ☐ { 0 ____ − ____ ; 6 ____ − ____ }

# Lesson 9

Facts + Bonus = TOTAL

**1**

| A | B | C | D | E |
|---|---|---|---|---|
| 4285 | 3624 | 8153 | 9284 | 7295 |

**2**

$$15 - 10 \qquad 12 - 10 \qquad 14 - 10 \qquad 11 - 10 \qquad 13 - 10 \qquad 15 - 10 \qquad 12 - 10 \qquad 11 - 10$$

**3**

$$8 - 0 \qquad 4 - 1 \qquad 7 - 0 \qquad 3 - 0 \qquad 10 - 1 \qquad 5 - 0 \qquad 8 - 1 \qquad 10 - 0$$

**4**

**A** 14 { 7, 7 — ____ + ____ ; ____ − ____

**B** 16 { 8, 8 — ____ + ____ ; ____ − ____

**C** 18 { 9, 9 — ____ + ____ ; ____ − ____

**D** 12 { 6, 6 — ____ + ____ ; ____ − ____

**5**

$$14 - 7 \qquad 10 - 1 \qquad 18 - 9 \qquad 8 - 4 \qquad 7 - 0 \qquad 10 - 5 \qquad 6 - 0 \qquad 6 - 3$$

$$10 - 0 \qquad 18 - 9 \qquad 10 - 5 \qquad 8 - 4 \qquad 8 - 0 \qquad 14 - 7 \qquad 6 - 3 \qquad 5 - 1$$

Part 5 continues on the next page.

| 8 | 6 | 14 | 3 | 18 | 8 | 6 | 6 |
|---|---|---|---|---|---|---|---|
| − 4 | − 1 | − 7 | − 0 | − 9 | − 1 | − 3 | − 0 |

| 9 | 9 | 6 | 6 | 6 | 8 | 5 | 8 |
|---|---|---|---|---|---|---|---|
| − 1 | − 0 | − 3 | − 1 | − 0 | − 4 | − 1 | − 0 |

## 6

**A**
1 ______ − ______
2 ______ − ______

**B**
0 ______ − ______
7 ______ − ______

**C**
0 ______ − ______
4 ______ − ______

**D**
1 ______ − ______
8 ______ − ______

**E**
0 ______ − ______
9 ______ − ______

**F**
1 ______ − ______
6 ______ − ______

**G**
1 ______ − ______
4 ______ − ______

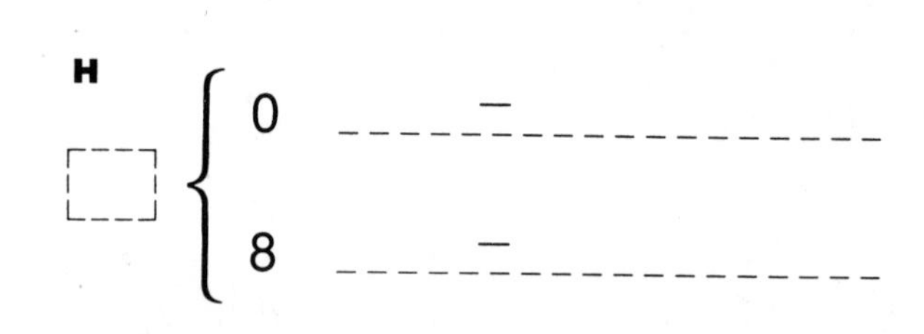

**H**
0 ______ − ______
8 ______ − ______

# Lesson 10

| Facts | + | Bonus | = | TOTAL |
|---|---|---|---|---|
| | | | | |

## 1

| A | B | C | D | E |
|---|---|---|---|---|
| 3279 | 146 | 6571 | 432 | 2384 |

## 2

$14 - 10$ $17 - 10$ $13 - 10$ $15 - 10$ $18 - 10$ $12 - 10$ $16 - 10$ $19 - 10$

## 3

$7 - 0$ $9 - 0$ $4 - 1$ $8 - 0$ $3 - 1$ $6 - 0$ $10 - 1$ $10 - 0$

## 4

**A** 18: 9 + 9 ____; 9 − 9 ____

**B** 8: 4 + 4 ____; 4 − 4 ____

**C** 10: 5 + 5 ____; 5 − 5 ____

**D** 6: 3 + 3 ____; 3 − 3 ____

**E** 12: 6 + 6 ____; 6 − 6 ____

**F** 14: 7 + 7 ____; 7 − 7 ____

## 5

$14 - 7$ $10 - 5$ $18 - 9$ $8 - 4$ $6 - 3$ $14 - 7$ $10 - 5$ $18 - 9$

## 6

| 18 − 9 | 14 − 7 | 10 − 5 | 8 − 0 | 6 − 3 | 8 − 4 | 18 − 9 | 14 − 7 |
|---|---|---|---|---|---|---|---|
| 7 − 0 | 5 − 1 | 14 − 7 | 10 − 1 | 10 − 0 | 10 − 5 | 8 − 1 | 8 − 4 |
| 8 − 0 | 18 − 9 | 8 − 4 | 10 − 5 | 14 − 7 | 5 − 0 | 7 − 0 | 3 − 1 |
| 5 − 1 | 8 − 4 | 7 − 0 | 9 − 1 | 4 − 2 | 18 − 9 | 14 − 7 | 8 − 4 |

## 7

**A**
- 0 ___ − ___
- 7 ___ − ___

**B**
- 1 ___ − ___
- 3 ___ − ___

**C**
- 0 ___ − ___
- 5 ___ − ___

**D**
- 1 ___ − ___
- 8 ___ − ___

**E**
- 0 ___ − ___
- 4 ___ − ___

**F**
- 1 ___ − ___
- 7 ___ − ___

# Lesson 11

Facts + Problems + Bonus = TOTAL

## 1

| A | B | C | D | E | F |
|---|---|---|---|---|---|
| 3856 | 4258 | 326 | 1420 | 350 | 4263 |

## 2

| 14 | 18 | 15 | 19 | 13 | 17 | 12 | 16 |
|---|---|---|---|---|---|---|---|
| − 10 | − 10 | − 10 | − 10 | − 10 | − 10 | − 10 | − 10 |

## 3

**A** 18 { 9 ___ + ___ ; 9 ___ − ___ }

**B** 14 { 7 ___ + ___ ; 7 ___ − ___ }

**C** 16 { 8 ___ + ___ ; 8 ___ − ___ }

**D** 10 { 5 ___ + ___ ; 5 ___ − ___ }

## 4

| 18 | 14 | 10 | 10 | 8 | 8 | 14 | 18 |
|---|---|---|---|---|---|---|---|
| − 9 | − 7 | − 5 | − 1 | − 0 | − 4 | − 7 | − 9 |

| 9 | 9 | 6 | 5 | 10 | 6 | 18 | 7 |
|---|---|---|---|---|---|---|---|
| − 0 | − 1 | − 3 | − 1 | − 5 | − 1 | − 9 | − 0 |

| 18 | 7 | 8 | 3 | 18 | 8 | 8 | 14 |
|---|---|---|---|---|---|---|---|
| − 9 | − 0 | − 4 | − 0 | − 9 | − 1 | − 4 | − 7 |

## 5

**A**

☐ { 1 ___ – ___
6 ___ – ___

**B**

☐ { 0 ___ – ___
5 ___ – ___

**C**

☐ { 0 ___ – ___
4 ___ – ___

**D**

☐ { 1 ___ – ___
9 ___ – ___

## 6

$\begin{array}{r}10\\-10\\\hline\end{array}$ $\begin{array}{r}9\\-8\\\hline\end{array}$ $\begin{array}{r}7\\-6\\\hline\end{array}$ $\begin{array}{r}4\\-4\\\hline\end{array}$ $\begin{array}{r}8\\-7\\\hline\end{array}$ $\begin{array}{r}5\\-5\\\hline\end{array}$ $\begin{array}{r}9\\-9\\\hline\end{array}$ $\begin{array}{r}10\\-9\\\hline\end{array}$

$\begin{array}{r}5\\-4\\\hline\end{array}$ $\begin{array}{r}3\\-3\\\hline\end{array}$ $\begin{array}{r}8\\-8\\\hline\end{array}$ $\begin{array}{r}3\\-2\\\hline\end{array}$ $\begin{array}{r}5\\-4\\\hline\end{array}$ $\begin{array}{r}7\\-7\\\hline\end{array}$ $\begin{array}{r}2\\-2\\\hline\end{array}$ $\begin{array}{r}6\\-5\\\hline\end{array}$

## 7

**A** $\overset{3}{\cancel{4}}{}^{1}59$

**B** $\cancel{7}4$

**C** $\cancel{3}572$

**D** $\cancel{8}40$

## 8

**A** $\begin{array}{r}841\\-410\\\hline\end{array}$

**B** $\begin{array}{r}537\\-101\\\hline\end{array}$

**C** $\begin{array}{r}328\\-104\\\hline\end{array}$

## 9

**A** $\begin{array}{r}546\\-101\\\hline\end{array}$

**B** $\begin{array}{r}546\\+101\\\hline\end{array}$

**C** $\begin{array}{r}4863\\-2100\\\hline\end{array}$

**D** $\begin{array}{r}39\\-11\\\hline\end{array}$

**E** $\begin{array}{r}5263\\+1102\\\hline\end{array}$

# Lesson 12

| Test | + | Facts | + | Problems | + | Bonus | = | TOTAL |
|---|---|---|---|---|---|---|---|---|
| | | | | | | | | |

## 1

$$\begin{array}{r}15\\-10\\\hline\end{array}\quad\begin{array}{r}18\\-10\\\hline\end{array}\quad\begin{array}{r}13\\-10\\\hline\end{array}\quad\begin{array}{r}17\\-10\\\hline\end{array}\quad\begin{array}{r}15\\-10\\\hline\end{array}\quad\begin{array}{r}12\\-10\\\hline\end{array}\quad\begin{array}{r}14\\-10\\\hline\end{array}\quad\begin{array}{r}19\\-10\\\hline\end{array}$$

## 2

**A** 4236 **B** 8150 **C** 425 **D** 7111 **E** 711 **F** 250

## 3

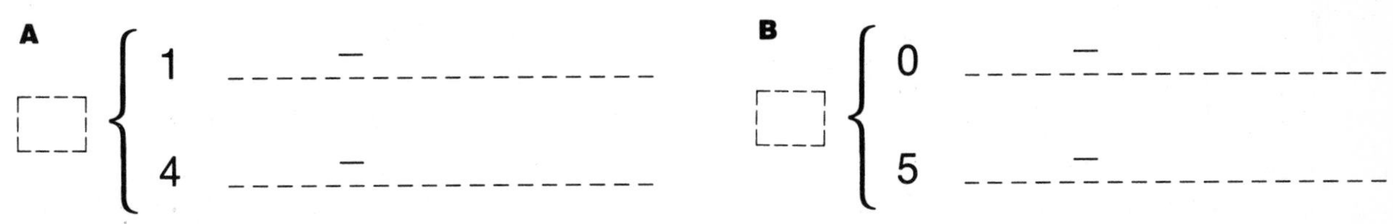

**A** { 1 ______ – ______ ; 4 ______ – ______ }

**B** { 0 ______ – ______ ; 5 ______ – ______ }

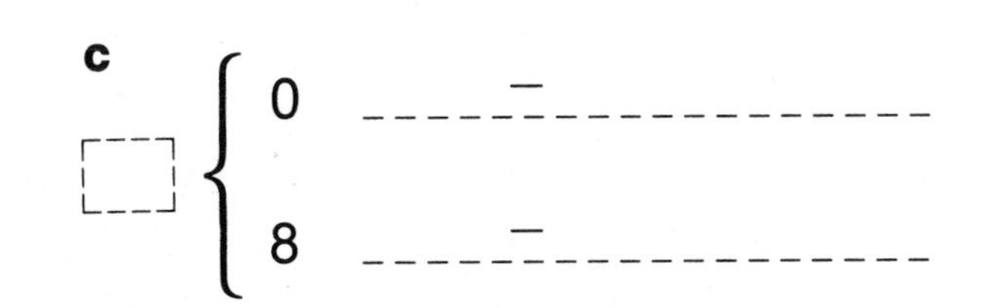

**C** { 0 ______ – ______ ; 8 ______ – ______ }

**D** { 1 ______ – ______ ; 9 ______ – ______ }

**E** { 1 ______ – ______ ; 7 ______ – ______ }

**F** { 0 ______ – ______ ; 4 ______ – ______ }

## 4

$$\begin{array}{r}10\\-10\\\hline\end{array}\quad\begin{array}{r}4\\-3\\\hline\end{array}\quad\begin{array}{r}14\\-6\\\hline\end{array}\quad\begin{array}{r}8\\-7\\\hline\end{array}\quad\begin{array}{r}14\\-8\\\hline\end{array}\quad\begin{array}{r}2\\-2\\\hline\end{array}\quad\begin{array}{r}9\\-8\\\hline\end{array}\quad\begin{array}{r}5\\-4\\\hline\end{array}$$

$$\begin{array}{r}10\\-9\\\hline\end{array}\quad\begin{array}{r}5\\-5\\\hline\end{array}\quad\begin{array}{r}3\\-2\\\hline\end{array}\quad\begin{array}{r}14\\-8\\\hline\end{array}\quad\begin{array}{r}4\\-4\\\hline\end{array}\quad\begin{array}{r}14\\-6\\\hline\end{array}\quad\begin{array}{r}3\\-2\\\hline\end{array}\quad\begin{array}{r}6\\-5\\\hline\end{array}$$

## 5

$$\begin{array}{r} 8 \\ -\ 0 \\ \hline \end{array} \quad \begin{array}{r} 4 \\ -\ 0 \\ \hline \end{array} \quad \begin{array}{r} 7 \\ -\ 1 \\ \hline \end{array} \quad \begin{array}{r} 8 \\ -\ 0 \\ \hline \end{array} \quad \begin{array}{r} 3 \\ -\ 1 \\ \hline \end{array} \quad \begin{array}{r} 9 \\ -\ 0 \\ \hline \end{array} \quad \begin{array}{r} 5 \\ -\ 0 \\ \hline \end{array} \quad \begin{array}{r} 8 \\ -\ 1 \\ \hline \end{array}$$

## 6

**A** $6\overset{3}{\cancel{4}}{}^{1}8$

**B** $37\cancel{9}0$

**C** $\cancel{7}354$

**D** $\cancel{1}48$

**E** $5\cancel{9}35$

## 7

**A**
$$\begin{array}{r} 496 \\ -\ 200 \\ \hline \end{array}$$

**B**
$$\begin{array}{r} 638 \\ -\ 104 \\ \hline \end{array}$$

**C**
$$\begin{array}{r} 345 \\ -\ 121 \\ \hline \end{array}$$

## 8

**A**
$$\begin{array}{r} 985 \\ -\ 101 \\ \hline \end{array}$$

**B**
$$\begin{array}{r} 985 \\ +\ 901 \\ \hline \end{array}$$

**C**
$$\begin{array}{r} 452 \\ +\ 201 \\ \hline \end{array}$$

**D**
$$\begin{array}{r} 488 \\ -\ 101 \\ \hline \end{array}$$

**E**
$$\begin{array}{r} 472 \\ -\ 200 \\ \hline \end{array}$$

**F**
$$\begin{array}{r} 759 \\ +\ 110 \\ \hline \end{array}$$

**G**
$$\begin{array}{r} 5263 \\ +\ 1102 \\ \hline \end{array}$$

**H**
$$\begin{array}{r} 485 \\ -\ 120 \\ \hline \end{array}$$

# Lesson 13

| Facts | + | Problems | + | Bonus | = | TOTAL |
|---|---|---|---|---|---|---|
| | | | | | | |

## 1

A 4000 B 400 C 318 D 1150 E 2314 F 107

## 2

$14 - 8$ $14 - 10$ $14 - 6$ $14 - 7$ $14 - 8$ $14 - 10$ $14 - 6$ $14 - 7$

## 3

$12 - 3$ $12 - 4$ $12 - 5$ $12 - 4$ $12 - 3$ $12 - 5$ $12 - 4$ $12 - 5$

## 4

A { 7 ___ – ___ ; 1 ___ – ___ }

B { 9 ___ – ___ ; 0 ___ – ___ }

C { 4 ___ – ___ ; 0 ___ – ___ }

D { 5 ___ – ___ ; 1 ___ – ___ }

## 5

$8 - 8$ $6 - 5$ $4 - 3$ $7 - 7$ $3 - 3$ $9 - 8$ $10 - 9$ $8 - 8$

$9 - 9$ $7 - 6$ $5 - 5$ $8 - 7$ $3 - 2$ $6 - 6$ $10 - 10$ $10 - 9$

## 6

**A**

$$12 \begin{cases} 6 \quad \text{\_\_\_\_} + \text{\_\_\_\_} \\ 6 \quad \text{\_\_\_\_} - \text{\_\_\_\_} \end{cases}$$

**B**

$$\square \begin{cases} 7 \quad \text{\_\_\_\_} + \text{\_\_\_\_} \\ 7 \quad \text{\_\_\_\_} - \text{\_\_\_\_} \end{cases}$$

**C**

$$\square \begin{cases} 9 \quad \text{\_\_\_\_} + \text{\_\_\_\_} \\ 9 \quad \text{\_\_\_\_} - \text{\_\_\_\_} \end{cases}$$

**D**

$$\square \begin{cases} 8 \quad \text{\_\_\_\_} + \text{\_\_\_\_} \\ 8 \quad \text{\_\_\_\_} - \text{\_\_\_\_} \end{cases}$$

## 7

$\begin{array}{r} 12 \\ -\ 6 \\ \hline \end{array}$ $\begin{array}{r} 11 \\ -10 \\ \hline \end{array}$ $\begin{array}{r} 14 \\ -\ 7 \\ \hline \end{array}$ $\begin{array}{r} 15 \\ -10 \\ \hline \end{array}$ $\begin{array}{r} 16 \\ -\ 8 \\ \hline \end{array}$ $\begin{array}{r} 13 \\ -10 \\ \hline \end{array}$ $\begin{array}{r} 14 \\ -\ 7 \\ \hline \end{array}$ $\begin{array}{r} 19 \\ -10 \\ \hline \end{array}$

$\begin{array}{r} 18 \\ -\ 9 \\ \hline \end{array}$ $\begin{array}{r} 14 \\ -10 \\ \hline \end{array}$ $\begin{array}{r} 18 \\ -10 \\ \hline \end{array}$ $\begin{array}{r} 12 \\ -\ 6 \\ \hline \end{array}$ $\begin{array}{r} 8 \\ -4 \\ \hline \end{array}$ $\begin{array}{r} 8 \\ -0 \\ \hline \end{array}$ $\begin{array}{r} 16 \\ -\ 8 \\ \hline \end{array}$ $\begin{array}{r} 12 \\ -10 \\ \hline \end{array}$

$\begin{array}{r} 12 \\ -\ 6 \\ \hline \end{array}$ $\begin{array}{r} 14 \\ -\ 7 \\ \hline \end{array}$ $\begin{array}{r} 16 \\ -10 \\ \hline \end{array}$ $\begin{array}{r} 16 \\ -\ 8 \\ \hline \end{array}$ $\begin{array}{r} 18 \\ -\ 9 \\ \hline \end{array}$ $\begin{array}{r} 6 \\ -3 \\ \hline \end{array}$ $\begin{array}{r} 6 \\ -1 \\ \hline \end{array}$ $\begin{array}{r} 10 \\ -\ 1 \\ \hline \end{array}$

## 8

**A** 2~~1~~7 **B** 7~~2~~04 **C** 3~~1~~2 **D** ~~1~~730 **E** ~~7~~7 **F** 3~~4~~81

## 9

**A** $\begin{array}{r} 348 \\ -\quad 7 \\ \hline \end{array}$ **B** $\begin{array}{r} 4325 \\ -\quad 10 \\ \hline \end{array}$ **C** $\begin{array}{r} 5347 \\ -\quad 31 \\ \hline \end{array}$ **D** $\begin{array}{r} 268 \\ -\ 17 \\ \hline \end{array}$ **E** $\begin{array}{r} 64 \\ -\ 3 \\ \hline \end{array}$

## 10

**A** $\begin{array}{r} 643 \\ -510 \\ \hline \end{array}$ **B** $\begin{array}{r} 685 \\ +106 \\ \hline \end{array}$ **C** $\begin{array}{r} 3426 \\ +2183 \\ \hline \end{array}$ **D** $\begin{array}{r} 7356 \\ -1045 \\ \hline \end{array}$ **E** $\begin{array}{r} 758 \\ -111 \\ \hline \end{array}$

# Lesson 14

Facts + Problems + Bonus = TOTAL

## 1

| A | B | C | D | E | F |
|---|---|---|---|---|---|
| 508 | 3740 | 2851 | 1418 | 231 | 3266 |

## 2

**A** ☐ { 10 ___ – ___ ; 8 ___ – ___ }

**B** ☐ { 10 ___ – ___ ; 2 ___ – ___ }

**C** ☐ { 10 ___ – ___ ; 9 ___ – ___ }

**D** ☐ { 10 ___ – ___ ; 4 ___ – ___ }

## 3

$12 - 5$  $12 - 3$  $12 - 4$  $12 - 6$  $12 - 3$  $12 - 5$  $12 - 4$  $12 - 5$

$12 - 6$  $12 - 4$  $12 - 3$  $12 - 5$  $12 - 6$  $12 - 4$  $12 - 3$  $12 - 5$

## 4

$10 - 9$  $7 - 7$  $4 - 0$  $8 - 1$  $6 - 5$  $4 - 4$  $6 - 1$  $3 - 3$

$5 - 0$  $8 - 0$  $9 - 8$  $6 - 6$  $4 - 0$  $7 - 1$  $4 - 3$  $5 - 3$

## 5

$16 - 10$  $16 - 8$  $14 - 7$  $14 - 10$  $14 - 8$  $14 - 6$  $18 - 9$  $12 - 6$

Part 5 continues on the next page.

| | | | | | | | |
|---|---|---|---|---|---|---|---|
| 12 − 10 | 16 − 8 | 14 − 6 | 14 − 8 | 12 − 6 | 15 − 10 | 14 − 8 | 10 − 5 |
| 14 − 6 | 14 − 10 | 14 − 7 | 14 − 8 | 8 − 4 | 6 − 3 | 18 − 9 | 16 − 8 |

## 6

A. 7̸3

B. 91̸4

C. 3̸74

D. 1̸380

E. 21̸04

F. 53̸25

## 7

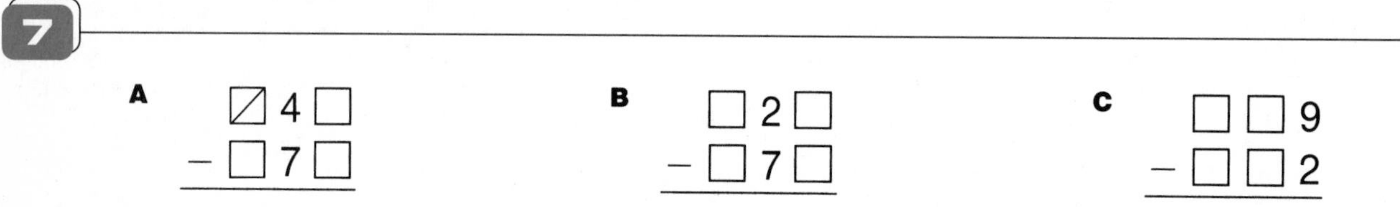

A. □4□ − □7□

B. □2□ − □7□

C. □□9 − □□2

D. □□6□ − □7□

E. □□9 − □□8

F. □3□□ − □4□□

G. □□2 − □3

H. □□5□ − □3□

## 8

A. 937 − 1

B. 469 − 8

C. 6284 − 83

D. 4321 − 10

E. 5374 − 62

## 9

A. 436 + 216

B. 443 − 201

C. 734 − 622

D. 568 + 128

E. 364 − 132

# Lesson 15

| Test | Facts | Problems | Bonus | TOTAL |
|---|---|---|---|---|
| | | | | |

## 1

| | | | | | | | |
|---|---|---|---|---|---|---|---|
| 12 − 3 | 12 − 6 | 12 − 4 | 12 − 5 | 12 − 6 | 12 − 3 | 12 − 5 | 12 − 4 |
| 12 − 6 | 12 − 3 | 12 − 5 | 12 − 4 | 12 − 3 | 12 − 6 | 12 − 5 | 12 − 4 |

## 2

| | | | | | | | |
|---|---|---|---|---|---|---|---|
| 14 − 6 | 14 − 10 | 14 − 8 | 9 − 8 | 7 − 7 | 2 − 0 | 8 − 7 | 6 − 1 |
| 6 − 5 | 9 − 0 | 5 − 4 | 6 − 1 | 14 − 8 | 6 − 6 | 8 − 0 | 14 − 6 |
| 8 − 7 | 5 − 1 | 14 − 8 | 10 − 9 | 4 − 0 | 14 − 6 | 9 − 8 | 7 − 1 |

## 3

**A** { 10 ______ / 7 ______ }

**B** { 10 ______ / 9 ______ }

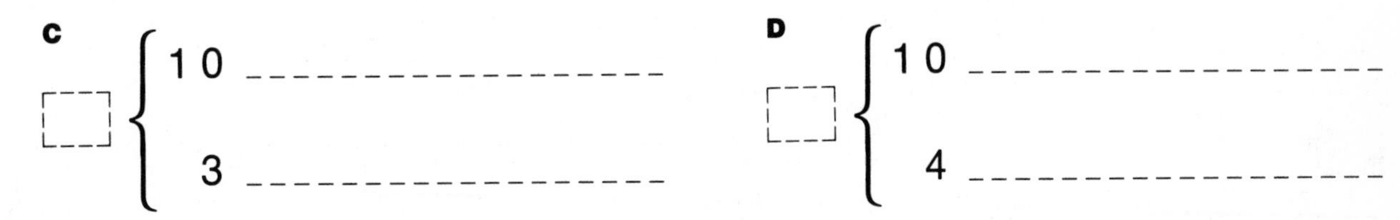


## 4

| A | B | C | D | E | F |
|---|---|---|---|---|---|
| 7214 | 9382 | 451 | 4160 | 380 | 5246 |

## 5

$$\begin{array}{r}7\\-1\\\hline\end{array}\quad\begin{array}{r}7\\-3\\\hline\end{array}\quad\begin{array}{r}7\\-2\\\hline\end{array}\quad\begin{array}{r}7\\-0\\\hline\end{array}\quad\begin{array}{r}7\\-3\\\hline\end{array}\quad\begin{array}{r}7\\-2\\\hline\end{array}\quad\begin{array}{r}7\\-1\\\hline\end{array}\quad\begin{array}{r}7\\-3\\\hline\end{array}$$

## 6

**A** $51\not{3}4$

**B** $\not{2}0$

**C** $6\not{3}0$

**D** $\not{7}82$

**E** $8\not{1}72$

**F** $9\not{1}02$

## 7

**A**
```
  □ ⧄ 3 □
− □ □ 5 □
---------
```

**B**
```
  □ □ 4 □
− □ □ 6 □
---------
```

**C**
```
  □ □ 8
− □ □ 8
-------
```

**D**
```
  □ □ 6 □
− □ □ 7 □
---------
```

**E**
```
  □ 0 □ □
− □ 5 □ □
---------
```

**F**
```
  □ □ 3 □
−   □ 1 □
---------
```

**G**
```
  □ □ 2
− □ □ 2
-------
```

**H**
```
  □ □ 7
− □ □ 8
-------
```

## 8

$$\begin{array}{r}12\\-\ 6\\\hline\end{array}\quad\begin{array}{r}14\\-\ 7\\\hline\end{array}\quad\begin{array}{r}10\\-\ 5\\\hline\end{array}\quad\begin{array}{r}12\\-\ 6\\\hline\end{array}\quad\begin{array}{r}18\\-\ 9\\\hline\end{array}\quad\begin{array}{r}8\\-4\\\hline\end{array}\quad\begin{array}{r}6\\-3\\\hline\end{array}\quad\begin{array}{r}16\\-\ 8\\\hline\end{array}$$

$$\begin{array}{r}18\\-10\\\hline\end{array}\quad\begin{array}{r}14\\-10\\\hline\end{array}\quad\begin{array}{r}14\\-\ 7\\\hline\end{array}\quad\begin{array}{r}16\\-10\\\hline\end{array}\quad\begin{array}{r}16\\-\ 8\\\hline\end{array}\quad\begin{array}{r}12\\-\ 6\\\hline\end{array}\quad\begin{array}{r}12\\-10\\\hline\end{array}\quad\begin{array}{r}18\\-\ 9\\\hline\end{array}$$

## 9

**A** $\begin{array}{r}983\\-\ 41\\\hline\end{array}$

**B** $\begin{array}{r}6928\\-\ 17\\\hline\end{array}$

**C** $\begin{array}{r}626\\-\ 13\\\hline\end{array}$

**D** $\begin{array}{r}2648\\-\ 324\\\hline\end{array}$

**E** $\begin{array}{r}46\\-\ 5\\\hline\end{array}$

## 10

**A** $\begin{array}{r}9679\\-8370\\\hline\end{array}$

**B** $\begin{array}{r}4762\\+4610\\\hline\end{array}$

**C** $\begin{array}{r}868\\-467\\\hline\end{array}$

**D** $\begin{array}{r}4682\\-2502\\\hline\end{array}$

**E** $\begin{array}{r}4438\\+2028\\\hline\end{array}$

# Lesson 16

| Facts | + | Problems | + | Bonus | = | TOTAL |
|---|---|---|---|---|---|---|
| | | | | | | |

## 1

| A | B | C | D | E |
|---|---|---|---|---|
| 4318 | 6204 | 704 | 1824 | 314 |

## 2

$7 - 3$   $7 - 1$   $7 - 2$   $7 - 0$   $7 - 2$   $7 - 3$   $7 - 1$   $7 - 3$

## 3

$12 - 4$   $12 - 5$   $12 - 6$   $12 - 3$   $12 - 5$   $12 - 4$   $12 - 6$   $12 - 3$

$12 - 5$   $12 - 3$   $12 - 6$   $12 - 4$   $12 - 5$   $12 - 3$   $12 - 6$   $12 - 5$

## 4

$16 - 10$   $16 - 8$   $14 - 6$   $14 - 10$   $14 - 7$   $14 - 8$   $12 - 10$   $12 - 6$

$18 - 10$   $14 - 6$   $8 - 7$   $10 - 5$   $10 - 9$   $8 - 4$   $8 - 1$   $8 - 8$

$9 - 8$   $7 - 7$   $6 - 0$   $7 - 1$   $14 - 8$   $5 - 0$   $8 - 7$   $14 - 6$

## 5

**A** □□4 − □□6

**B** □□3 − □□0

**C** □□6□ − □□9□

**D** □5 − □5

**E** □6 − □7

**F** □□9□ − □□8□

**G** □0 − □2

**H** □□8 − □□7

## 6

A
$$\begin{array}{r} \overset{2}{\cancel{3}}\,{}^{1}4 \\ -\ 18 \\ \hline \end{array}$$

B
$$\begin{array}{r} \overset{7}{\cancel{8}}\,{}^{1}45 \\ -\ 371 \\ \hline \end{array}$$

C
$$\begin{array}{r} 7\overset{4}{\cancel{5}}\,{}^{1}03 \\ -\ 6212 \\ \hline \end{array}$$

D
$$\begin{array}{r} \overset{2}{\cancel{3}}\,{}^{1}685 \\ -\ 1814 \\ \hline \end{array}$$

## 7

A
$$\begin{array}{r} 52 \\ -\ 36 \\ \hline \end{array}$$

B
$$\begin{array}{r} 944 \\ -\ 173 \\ \hline \end{array}$$

C
$$\begin{array}{r} 260 \\ -\ 155 \\ \hline \end{array}$$

D
$$\begin{array}{r} 824 \\ -\ 461 \\ \hline \end{array}$$

E
$$\begin{array}{r} 2876 \\ -\ 1458 \\ \hline \end{array}$$

## 8

A
$$\begin{array}{r} 387 \\ +\ \ \ 7 \\ \hline \end{array}$$

B
$$\begin{array}{r} 4258 \\ -\ \ 114 \\ \hline \end{array}$$

C
$$\begin{array}{r} 3686 \\ -\ \ 513 \\ \hline \end{array}$$

D
$$\begin{array}{r} 508 \\ +\ \ \ 8 \\ \hline \end{array}$$

E
$$\begin{array}{r} 4378 \\ -\ \ \ 68 \\ \hline \end{array}$$

F
$$\begin{array}{r} 4216 \\ -\ \ 100 \\ \hline \end{array}$$

G
$$\begin{array}{r} 3824 \\ +\ 1810 \\ \hline \end{array}$$

H
$$\begin{array}{r} 7538 \\ -\ \ 430 \\ \hline \end{array}$$

I
$$\begin{array}{r} 426 \\ -\ \ 20 \\ \hline \end{array}$$

J
$$\begin{array}{r} 653 \\ +\ 253 \\ \hline \end{array}$$

# Lesson 17

| Facts | + | Problems | + | Bonus | = | TOTAL |
|---|---|---|---|---|---|---|
| | | | | | | |

## 1

| 7 | 7 | 7 | 7 | 7 | 7 | 7 | 7 |
|---|---|---|---|---|---|---|---|
| − 3 | − 0 | − 2 | − 1 | − 3 | − 1 | − 2 | − 3 |

## 2

**A** 3406

**B** 182

**C** 5414

**D** 3811

**E** 701

## 3

**A** 12 { 5, ☐

**B** 12 { 4, ☐

**C** 12 { 3, ☐

**D** 12 { 2, ☐

## 4

| 16 | 16 | 18 | 14 | 18 | 12 | 8 | 6 |
|---|---|---|---|---|---|---|---|
| − 8 | − 10 | − 9 | − 7 | − 10 | − 6 | − 4 | − 3 |

| 18 | 8 | 14 | 14 | 16 | 10 | 10 | 16 |
|---|---|---|---|---|---|---|---|
| − 9 | − 7 | − 7 | − 10 | − 8 | − 9 | − 5 | − 8 |

## 5

| 12 | 14 | 12 | 12 | 8 | 12 | 14 | 14 |
|---|---|---|---|---|---|---|---|
| − 5 | − 6 | − 6 | − 4 | − 8 | − 3 | − 7 | − 8 |

| 7 | 12 | 14 | 12 | 5 | 9 | 12 | 16 |
|---|---|---|---|---|---|---|---|
| − 0 | − 4 | − 6 | − 5 | − 4 | − 0 | − 3 | − 8 |

Part 5 continues on the next page.

$\begin{array}{r} 8 \\ -7 \\ \hline \end{array}$ $\begin{array}{r} 12 \\ -\ 5 \\ \hline \end{array}$ $\begin{array}{r} 18 \\ -10 \\ \hline \end{array}$ $\begin{array}{r} 10 \\ -\ 1 \\ \hline \end{array}$ $\begin{array}{r} 12 \\ -\ 3 \\ \hline \end{array}$ $\begin{array}{r} 18 \\ -\ 9 \\ \hline \end{array}$ $\begin{array}{r} 14 \\ -\ 8 \\ \hline \end{array}$ $\begin{array}{r} 12 \\ -\ 4 \\ \hline \end{array}$

## 6

**A** $\begin{array}{r} \square\square5\square \\ -\square\square7\square \\ \hline \end{array}$ **B** $\begin{array}{r} \square\square5\square \\ -\square\square5\square \\ \hline \end{array}$ **C** $\begin{array}{r} \square8\square\square \\ -\square7\square\square \\ \hline \end{array}$ **D** $\begin{array}{r} \square5 \\ -\square0 \\ \hline \end{array}$

**E** $\begin{array}{r} \square0\square \\ -\square8\square \\ \hline \end{array}$ **F** $\begin{array}{r} \square\square7 \\ -\square\square7 \\ \hline \end{array}$ **G** $\begin{array}{r} \square\square3 \\ -\square\square8 \\ \hline \end{array}$ **H** $\begin{array}{r} \square4\square\square \\ -\square0\square\square \\ \hline \end{array}$

## 7

**A** $\begin{array}{r} \overset{5}{\cancel{6}}{}^{1}2 \\ -16 \\ \hline \end{array}$ **B** $\begin{array}{r} 65\overset{2}{\cancel{3}}{}^{1}4 \\ -3417 \\ \hline \end{array}$ **C** $\begin{array}{r} 4\overset{2}{\cancel{3}}{}^{1}85 \\ -1195 \\ \hline \end{array}$ **D** $\begin{array}{r} \overset{2}{\cancel{3}}{}^{1}453 \\ -1641 \\ \hline \end{array}$

## 8

**A** $\begin{array}{r} 42 \\ -26 \\ \hline \end{array}$ **B** $\begin{array}{r} 704 \\ -153 \\ \hline \end{array}$ **C** $\begin{array}{r} 6284 \\ -5017 \\ \hline \end{array}$ **D** $\begin{array}{r} 3648 \\ -1807 \\ \hline \end{array}$ **E** $\begin{array}{r} 2863 \\ -1681 \\ \hline \end{array}$

## 9

**A** $\begin{array}{r} 749 \\ +109 \\ \hline \end{array}$ **B** $\begin{array}{r} 2371 \\ -\ \ 70 \\ \hline \end{array}$ **C** $\begin{array}{r} 86 \\ +26 \\ \hline \end{array}$ **D** $\begin{array}{r} 5714 \\ -1602 \\ \hline \end{array}$ **E** $\begin{array}{r} 3428 \\ -\ \ 24 \\ \hline \end{array}$

**F** $\begin{array}{r} 3864 \\ -\ 732 \\ \hline \end{array}$ **G** $\begin{array}{r} 5624 \\ +\ 635 \\ \hline \end{array}$ **H** $\begin{array}{r} 408 \\ -\ \ 4 \\ \hline \end{array}$ **I** $\begin{array}{r} 6298 \\ -\ \ 88 \\ \hline \end{array}$ **J** $\begin{array}{r} 4582 \\ +\ 182 \\ \hline \end{array}$

# Lesson 18

Facts + Problems + Bonus = TOTAL

## 1

| | | | | | | | |
|---|---|---|---|---|---|---|---|
| 12 − 4 | 7 − 2 | 7 − 3 | 12 − 3 | 12 − 5 | 7 − 1 | 12 − 3 | 12 − 4 |
| 7 − 0 | 12 − 5 | 7 − 2 | 12 − 3 | 7 − 3 | 12 − 4 | 12 − 5 | 12 − 3 |

## 2

**A** 12 { 4 ____ / ☐ ____

**B** 12 { 3 ____ / ☐ ____

**C** 12 { 5 ____ / ☐ ____

**D** 12 { 2 ____ / ☐ ____

## 3

| | | | | | | | |
|---|---|---|---|---|---|---|---|
| 12 − 7 | 12 − 9 | 12 − 8 | 12 − 6 | 12 − 8 | 12 − 9 | 12 − 7 | 12 − 6 |
| 12 − 4 | 12 − 8 | 12 − 5 | 12 − 7 | 12 − 3 | 12 − 9 | 12 − 6 | 12 − 5 |

## 4

**A** 5803

**B** 211

**C** 6414

**D** 814

**E** 9909

## 5

| | | | | | | | |
|---|---|---|---|---|---|---|---|
| 14 − 6 | 14 − 10 | 14 − 8 | 14 − 7 | 12 − 10 | 12 − 5 | 12 − 4 | 12 − 6 |

Part 5 continues on the next page.

$18 - 9$ | $9 - 0$ | $14 - 8$ | $16 - 10$ | $6 - 6$ | $15 - 10$ | $16 - 8$ | $8 - 7$

$12 - 10$ | $10 - 1$ | $8 - 4$ | $14 - 6$ | $3 - 0$ | $18 - 9$ | $12 - 6$ | $8 - 8$

$14 - 8$ | $14 - 7$ | $5 - 0$ | $13 - 10$ | $5 - 5$ | $16 - 8$ | $12 - 4$ | $14 - 10$

## 6

**A** $396 - 248$

**B** $607 - 497$

**C** $848 - 670$

**D** $942 - 180$

**E** $7875 - 2965$

**F** $932 - 826$

**G** $3643 - 1840$

**H** $5990 - 4189$

**I** $825 - 160$

**J** $3046 - 1506$

## 7

**A** $8425 + 35$

**B** $4186 - 143$

**C** $386 + 180$

**D** $57 + 37$

**E** $386 - 43$

**F** $4163 - 2033$

**G** $865 - 455$

**H** $391 + 98$

**I** $3286 - 43$

**J** $528 + 8$

# Lesson 19

| Facts | + | Problems | + | Bonus | = | TOTAL |
|---|---|---|---|---|---|---|
| | | | | | | |

## 1

$7 - 3$   $7 - 0$   $12 - 3$   $7 - 1$   $12 - 5$   $12 - 4$   $7 - 2$   $12 - 3$

$7 - 3$   $12 - 4$   $7 - 1$   $12 - 5$   $7 - 2$   $12 - 5$   $12 - 3$   $12 - 4$

## 2

**A** 12 { 4 ____ ; ☐ ____

**B** 12 { 3 ____ ; ☐ ____

**C** 12 { 5 ____ ; ☐ ____

**D** 12 { 2 ____ ; ☐ ____

## 3

$12 - 8$   $12 - 6$   $12 - 9$   $12 - 7$   $12 - 9$   $12 - 7$   $12 - 8$   $12 - 6$

$12 - 5$   $12 - 9$   $12 - 4$   $12 - 3$   $12 - 7$   $12 - 8$   $12 - 5$   $12 - 9$

## 4

$14 - 6$   $8 - 4$   $14 - 10$   $9 - 8$   $14 - 8$   $12 - 6$   $18 - 9$   $16 - 10$

$10 - 5$   $14 - 6$   $6 - 0$   $18 - 9$   $14 - 8$   $15 - 10$   $9 - 9$   $17 - 10$

Part 4 continues on the next page.

| | | | | | | | |
|---|---|---|---|---|---|---|---|
| 14 | 14 | 18 | 6 | 11 | 6 | 14 | 14 |
| − 7 | − 8 | − 9 | − 5 | − 10 | − 3 | − 6 | − 7 |

| | | | | | | | |
|---|---|---|---|---|---|---|---|
| 18 | 18 | 13 | 8 | 15 | 6 | 16 | 13 |
| − 10 | − 9 | − 10 | − 4 | − 10 | − 5 | − 8 | − 10 |

## 5

| A | B | C | D | E |
|---|---|---|---|---|
| 3418 | 708 | 9210 | 5300 | 121 |

## 6

| A | B | C | D | E |
|---|---|---|---|---|
| 492 | 903 | 681 | 7043 | 76 |
| − 286 | − 791 | − 491 | − 5110 | − 38 |

| F | G | H | I | J |
|---|---|---|---|---|
| 3500 | 9601 | 6824 | 584 | 745 |
| − 1350 | − 490 | − 1904 | − 466 | − 80 |

## 7

| A | B | C | D | E |
|---|---|---|---|---|
| 8794 | 67 | 3625 | 646 | 6747 |
| − 1391 | − 36 | + 444 | + 108 | − 4601 |

| F | G | H | I | J |
|---|---|---|---|---|
| 3898 | 258 | 9148 | 3794 | 5234 |
| − 7 | + 8 | − 28 | + 96 | − 10 |

# Lesson 20

Facts + Problems + Bonus = TOTAL

## 1

| 10 | 10 | 10 | 10 | 10 | 10 | 10 | 10 |
|---|---|---|---|---|---|---|---|
| − 5 | − 4 | − 0 | − 1 | − 3 | − 2 | − 4 | − 2 |

| 10 | 10 | 10 | 10 | 10 | 10 | 10 | 10 |
|---|---|---|---|---|---|---|---|
| − 3 | − 4 | − 1 | − 5 | − 2 | − 3 | − 2 | − 4 |

## 2

**A**

12 { 4 ______ ; ☐ ______

**B**

12 { 2 ______ ; ☐ ______

**C**

12 { 3 ______ ; ☐ ______

**D**

12 { 5 ______ ; ☐ ______

## 3

| 12 | 12 | 12 | 12 | 12 | 12 | 12 | 12 |
|---|---|---|---|---|---|---|---|
| − 5 | − 8 | − 7 | − 9 | − 6 | − 7 | − 8 | − 9 |

| 12 | 12 | 12 | 12 | 12 | 12 | 12 | 12 |
|---|---|---|---|---|---|---|---|
| − 4 | − 9 | − 5 | − 8 | − 3 | − 7 | − 6 | − 10 |

## 4

| 7 | 7 | 14 | 8 | 14 | 5 | 7 | 14 |
|---|---|---|---|---|---|---|---|
| − 3 | − 2 | − 6 | − 0 | − 7 | − 5 | − 3 | − 8 |

| 10 | 18 | 6 | 18 | 14 | 12 | 16 | 9 |
|---|---|---|---|---|---|---|---|
| − 1 | − 9 | − 6 | − 10 | − 8 | − 10 | − 8 | − 8 |

Part 4 continues on the next page.

| | | | | | | | |
|---|---|---|---|---|---|---|---|
| 4 | 14 | 8 | 6 | 17 | 18 | 10 | 7 |
| − 0 | − 6 | − 4 | − 0 | − 10 | − 9 | − 9 | − 3 |

**5**

| A | B | C | D | E |
|---|---|---|---|---|
| 3204 | 814 | 1941 | 8111 | 414 |

**6**

| A | B | C | D |
|---|---|---|---|
| 3248 | 5082 | 9436 | 3410 |
| − 1639 | − 1966 | − 1728 | − 1609 |

**7**

| A | B | C | D | E |
|---|---|---|---|---|
| 824 | 7246 | 384 | 90 | 4603 |
| − 144 | − 138 | + 42 | − 49 | − 50 |

| F | G | H | I | J |
|---|---|---|---|---|
| 4327 | 830 | 4681 | 3415 | 7284 |
| + 1755 | − 105 | − 2871 | + 95 | − 92 |

| K | L | M | N | O |
|---|---|---|---|---|
| 3248 | 5700 | 480 | 648 | 286 |
| − 540 | − 10 | + 85 | − 180 | − 36 |

# Lesson 21

Facts + Problems + Bonus = TOTAL

## 1

$$\begin{array}{r}12\\-\ 9\\\hline\end{array}\quad\begin{array}{r}12\\-\ 5\\\hline\end{array}\quad\begin{array}{r}12\\-\ 6\\\hline\end{array}\quad\begin{array}{r}12\\-\ 8\\\hline\end{array}\quad\begin{array}{r}12\\-\ 7\\\hline\end{array}\quad\begin{array}{r}12\\-\ 4\\\hline\end{array}\quad\begin{array}{r}12\\-\ 3\\\hline\end{array}\quad\begin{array}{r}12\\-10\\\hline\end{array}$$

$$\begin{array}{r}12\\-\ 8\\\hline\end{array}\quad\begin{array}{r}12\\-\ 4\\\hline\end{array}\quad\begin{array}{r}12\\-\ 5\\\hline\end{array}\quad\begin{array}{r}12\\-\ 7\\\hline\end{array}\quad\begin{array}{r}12\\-\ 3\\\hline\end{array}\quad\begin{array}{r}12\\-\ 9\\\hline\end{array}\quad\begin{array}{r}12\\-\ 7\\\hline\end{array}\quad\begin{array}{r}12\\-\ 4\\\hline\end{array}$$

## 2

**A**

$$7\left\{\begin{array}{l}2\ \_\_\_\_\_\_\_\_\\\square\ \_\_\_\_\_\_\_\_\end{array}\right.$$

**B**

$$7\left\{\begin{array}{l}0\ \_\_\_\_\_\_\_\_\\\square\ \_\_\_\_\_\_\_\_\end{array}\right.$$

**C**

$$7\left\{\begin{array}{l}3\ \_\_\_\_\_\_\_\_\\\square\ \_\_\_\_\_\_\_\_\end{array}\right.$$

**D**

$$7\left\{\begin{array}{l}1\ \_\_\_\_\_\_\_\_\\\square\ \_\_\_\_\_\_\_\_\end{array}\right.$$

## 3

$$\begin{array}{r}14\\-\ 6\\\hline\end{array}\quad\begin{array}{r}19\\-10\\\hline\end{array}\quad\begin{array}{r}16\\-\ 8\\\hline\end{array}\quad\begin{array}{r}16\\-10\\\hline\end{array}\quad\begin{array}{r}14\\-\ 8\\\hline\end{array}\quad\begin{array}{r}18\\-\ 9\\\hline\end{array}\quad\begin{array}{r}12\\-10\\\hline\end{array}\quad\begin{array}{r}10\\-\ 5\\\hline\end{array}$$

$$\begin{array}{r}6\\-6\\\hline\end{array}\quad\begin{array}{r}18\\-\ 9\\\hline\end{array}\quad\begin{array}{r}10\\-\ 9\\\hline\end{array}\quad\begin{array}{r}14\\-\ 7\\\hline\end{array}\quad\begin{array}{r}16\\-10\\\hline\end{array}\quad\begin{array}{r}11\\-10\\\hline\end{array}\quad\begin{array}{r}10\\-\ 1\\\hline\end{array}\quad\begin{array}{r}8\\-7\\\hline\end{array}$$

## 4

$$\begin{array}{r}7\\-4\\\hline\end{array}\quad\begin{array}{r}7\\-5\\\hline\end{array}\quad\begin{array}{r}7\\-6\\\hline\end{array}\quad\begin{array}{r}7\\-2\\\hline\end{array}\quad\begin{array}{r}7\\-4\\\hline\end{array}\quad\begin{array}{r}7\\-3\\\hline\end{array}\quad\begin{array}{r}7\\-5\\\hline\end{array}\quad\begin{array}{r}7\\-7\\\hline\end{array}$$

## 5

**A** $\begin{array}{r}435\\-414\\\hline 21\end{array}$ **B** $\begin{array}{r}629\\-511\\\hline 18\end{array}$ **C** $\begin{array}{r}5734\\-5124\\\hline 610\end{array}$ **D** $\begin{array}{r}7765\\-7160\\\hline 605\end{array}$ **E** $\begin{array}{r}867\\-450\\\hline 17\end{array}$

**6**

A 3400

B 340

C 1824

D 4777

E 477

**7**

$10 - 4$

$10 - 3$

$10 - 1$

$10 - 2$

$10 - 5$

$10 - 3$

$10 - 4$

**8**

A $50 - 1 =$ ______

B $30 - 1 =$ ______

C $80 - 1 =$ ______

D $90 - 1 =$ ______

E $40 - 1 =$ ______

F $20 - 1 =$ ______

G $60 - 1 =$ ______

H $70 - 1 =$ ______

**9**

A $3420 - 1801$

B $8486 - 3768$

C $3092 - 1146$

D $3240 - 1739$

**10**

A $752 - 24$

B $3493 - 610$

C $3480 - 51$

D $70 + 38$

E $2400 - 90$

F $3528 + 528$

G $738 + 8$

H $9248 - 1068$

I $7251 - 601$

J $3846 + 3218$

K $460 - 355$

L $5284 + 6$

M $9328 - 16$

N $4822 - 142$

O $8605 - 7150$

# Lesson 22

| Test | + | Facts | + | Problems | + | Bonus | = | TOTAL |
|---|---|---|---|---|---|---|---|---|
| | | | | | | | | |

**1**

| 12 | 12 | 12 | 12 | 12 | 12 | 12 | 12 |
|---|---|---|---|---|---|---|---|
| − 7 | − 9 | − 6 | − 8 | − 5 | − 3 | − 8 | − 4 |

| 12 | 12 | 12 | 12 | 12 | 12 | 12 | 12 |
|---|---|---|---|---|---|---|---|
| − 7 | − 5 | − 4 | − 8 | − 6 | − 9 | − 3 | − 10 |

**2**

**A**
7 { 3 ______ ; [ ] ______

**B**
7 { 2 ______ ; [ ] ______

**C**
7 { 0 ______ ; [ ] ______

**D**
7 { 1 ______ ; [ ] ______

**3**

| 14 | 14 | 14 | 18 | 6 | 13 | 8 | 14 |
|---|---|---|---|---|---|---|---|
| − 10 | − 7 | − 8 | − 9 | − 6 | − 10 | − 0 | − 6 |

| 11 | 18 | 18 | 15 | 16 | 14 | 16 | 14 |
|---|---|---|---|---|---|---|---|
| − 10 | − 10 | − 9 | − 10 | − 10 | − 6 | − 8 | − 8 |

**4**

| A | B | C | D | E |
|---|---|---|---|---|
| 3261 | 5208 | 1418 | 555 | 5555 |

**5**

| 10 | 10 | 10 | 10 | 10 | 10 | 10 | 10 |
|---|---|---|---|---|---|---|---|
| − 4 | − 3 | − 2 | − 5 | − 1 | − 4 | − 2 | − 3 |

**6**

A
$$\begin{array}{r} 438 \\ -414 \\ \hline 24 \end{array}$$

B
$$\begin{array}{r} 329 \\ -211 \\ \hline 8 \end{array}$$

C
$$\begin{array}{r} 4834 \\ -4124 \\ \hline 10 \end{array}$$

D
$$\begin{array}{r} 7968 \\ -7160 \\ \hline 08 \end{array}$$

E
$$\begin{array}{r} 867 \\ -110 \\ \hline 7 \end{array}$$

**7**

A $60 - 1 =$ ______ B $70 - 1 =$ ______ C $20 - 1 =$ ______

D $40 - 1 =$ ______ E $30 - 1 =$ ______ F $50 - 1 =$ ______

G $90 - 1 =$ ______ H $80 - 1 =$ ______

**8**

A
$$\begin{array}{r} 3062 \\ -\ \ 954 \\ \hline \end{array}$$

B
$$\begin{array}{r} 3472 \\ -1855 \\ \hline \end{array}$$

C
$$\begin{array}{r} 4032 \\ -1515 \\ \hline \end{array}$$

D
$$\begin{array}{r} 3864 \\ -\ \ 956 \\ \hline \end{array}$$

**9**

A
$$\begin{array}{r} 3908 \\ -1718 \\ \hline \end{array}$$

B
$$\begin{array}{r} 840 \\ -280 \\ \hline \end{array}$$

C
$$\begin{array}{r} 596 \\ -490 \\ \hline \end{array}$$

D
$$\begin{array}{r} 70 \\ -35 \\ \hline \end{array}$$

E
$$\begin{array}{r} 5802 \\ -\ \ \ 91 \\ \hline \end{array}$$

F
$$\begin{array}{r} 4038 \\ -\ \ 510 \\ \hline \end{array}$$

G
$$\begin{array}{r} 3472 \\ -1455 \\ \hline \end{array}$$

H
$$\begin{array}{r} 785 \\ -640 \\ \hline \end{array}$$

# Lesson 23

Facts + Problems + Bonus = TOTAL

## 1

| | | | | | | | |
|---|---|---|---|---|---|---|---|
| $12 - 9$ | $12 - 4$ | $12 - 7$ | $12 - 8$ | $12 - 6$ | $12 - 3$ | $12 - 5$ | $12 - 4$ |
| $12 - 8$ | $12 - 6$ | $12 - 9$ | $12 - 7$ | $12 - 4$ | $12 - 8$ | $12 - 7$ | $12 - 3$ |

## 2

**A**

7 { 3 ________ / ☐ ________

**B**

7 { 1 ________ / ☐ ________

**C**

7 { 0 ________ / ☐ ________

**D**

7 { 2 ________ / ☐ ________

## 3

| | | | | | | | |
|---|---|---|---|---|---|---|---|
| $10 - 4$ | $10 - 3$ | $10 - 2$ | $10 - 5$ | $10 - 2$ | $10 - 4$ | $10 - 1$ | $10 - 5$ |
| $10 - 4$ | $10 - 2$ | $10 - 1$ | $10 - 3$ | $10 - 2$ | $10 - 4$ | $10 - 5$ | $10 - 0$ |

## 4

| | | | | | | | |
|---|---|---|---|---|---|---|---|
| $14 - 6$ | $12 - 10$ | $18 - 9$ | $18 - 10$ | $14 - 8$ | $15 - 10$ | $16 - 8$ | $10 - 5$ |
| $4 - 4$ | $18 - 9$ | $13 - 10$ | $8 - 4$ | $11 - 10$ | $8 - 7$ | $6 - 3$ | $8 - 0$ |

**5**

| 7 | 7 | 7 | 7 | 7 | 7 | 7 | 7 |
|---|---|---|---|---|---|---|---|
| − 5 | − 4 | − 6 | − 7 | − 4 | − 7 | − 6 | − 5 |

**6**

A 4056   B 5207   C 8024   D 2902   E 4908   F 7067

**7**

| A | B | C | D | E |
|---|---|---|---|---|
| 7986 | 918 | 4187 | 5288 | 9456 |
| − 6970 | − 914 | − 3106 | − 5274 | − 8413 |
| 16 | 4 | 81 | 14 | 43 |

**8**

A 50 − 1 = ______   B 80 − 1 = ______   C 70 − 1 = ______

D 20 − 1 = ______   E 60 − 1 = ______   F 40 − 1 = ______

G 30 − 1 = ______   H 90 − 1 = ______

**9**

| A | B | C | D | E |
|---|---|---|---|---|
| 9440 | 7030 | 943 | 9496 | 80 |
| − 4705 | − 3509 | + 971 | − 686 | − 25 |

| F | G | H | I | J |
|---|---|---|---|---|
| 3484 | 3420 | 4286 | 90 | 5290 |
| + 1086 | − 190 | − 196 | + 25 | − 3605 |

| K | L | M | N | O |
|---|---|---|---|---|
| 375 | 4034 | 248 | 4200 | 380 |
| − 340 | − 1526 | + 8 | − 90 | + 80 |

# Lesson 24

| Test | + | Facts | + | Problems | + | Bonus | = | TOTAL |
|---|---|---|---|---|---|---|---|---|
| | | | | | | | | |

## 1

| | | | | | | | |
|---|---|---|---|---|---|---|---|
| 10 | 10 | 10 | 10 | 10 | 10 | 10 | 10 |
| − 4 | − 2 | − 3 | − 1 | − 0 | − 2 | − 4 | − 5 |

| | | | | | | | |
|---|---|---|---|---|---|---|---|
| 10 | 10 | 10 | 10 | 10 | 10 | 10 | 10 |
| − 3 | − 1 | − 4 | − 2 | − 5 | − 3 | − 2 | − 4 |

## 2

**A**

7 { 1 ______ ; ☐ ______

**B**

7 { 3 ______ ; ☐ ______

**C**

7 { 0 ______ ; ☐ ______

**D**

7 { 2 ______ ; ☐ ______

## 3

| | | | | | | | |
|---|---|---|---|---|---|---|---|
| 14 | 14 | 5 | 5 | 18 | 6 | 14 | 10 |
| − 6 | − 7 | − 0 | − 4 | − 10 | − 0 | − 10 | − 9 |

| | | | | | | | |
|---|---|---|---|---|---|---|---|
| 16 | 13 | 7 | 14 | 7 | 8 | 8 | 15 |
| − 8 | − 10 | − 6 | − 8 | − 6 | − 7 | − 4 | − 10 |

| | | | | | | | |
|---|---|---|---|---|---|---|---|
| 10 | 14 | 18 | 8 | 17 | 8 | 14 | 12 |
| − 1 | − 6 | − 9 | − 8 | − 10 | − 0 | − 8 | − 6 |

## 4

**A** 4006

**B** 5027

**C** 7004

**D** 2902

**E** 4098

**F** 7007

## 5

| 12 | 12 | 12 | 12 | 12 | 12 | 12 | 12 |
|---|---|---|---|---|---|---|---|
| − 8 | − 7 | − 4 | − 6 | − 10 | − 5 | − 9 | − 3 |

| 12 | 12 | 12 | 12 | 12 | 12 | 12 | 12 |
|---|---|---|---|---|---|---|---|
| − 9 | − 10 | − 8 | − 7 | − 5 | − 3 | − 4 | − 6 |

## 6

| A | B | C | D | E |
|---|---|---|---|---|
| 5864 | 6993 | 5485 | 6274 | 3521 |
| − 4824 | − 6972 | − 1431 | − 6201 | − 2520 |
| 40 | 1 | 54 | 73 | 01 |

## 7

| A | B | C | D | E |
|---|---|---|---|---|
| 9958 | 5260 | 8356 | 3860 | 75 |
| − 9139 | − 3619 | + 156 | − 45 | + 70 |

| F | G | H | I | J |
|---|---|---|---|---|
| 3040 | 7410 | 396 | 6842 | 463 |
| − 931 | + 609 | − 8 | − 71 | − 80 |

| K | L | M | N | O |
|---|---|---|---|---|
| 4588 | 348 | 9480 | 3528 | 780 |
| + 28 | − 140 | − 1776 | − 360 | + 80 |

## 8

**A** 90 − 1 = ______ **B** 70 − 1 = ______ **C** 20 − 1 = ______

**D** 50 − 1 = ______ **E** 80 − 1 = ______ **F** 60 − 1 = ______

**G** 10 − 1 = ______ **H** 30 − 1 = ______

# Lesson 25

| Facts | + | Problems | + | Bonus | = | TOTAL |
|---|---|---|---|---|---|---|
| | | | | | | |

## 1

| 10 | 10 | 10 | 10 | 10 | 10 | 10 | 10 |
|---|---|---|---|---|---|---|---|
| − 4 | − 2 | − 3 | − 1 | − 4 | − 2 | − 3 | − 1 |

## 2

| 12 | 12 | 12 | 7 | 12 | 7 | 12 | 7 |
|---|---|---|---|---|---|---|---|
| − 5 | − 9 | − 8 | − 3 | − 7 | − 4 | − 9 | − 5 |

| 12 | 12 | 7 | 7 | 12 | 12 | 12 | 12 |
|---|---|---|---|---|---|---|---|
| − 8 | − 7 | − 4 | − 5 | − 9 | − 8 | − 7 | − 6 |

## 3

| 12 | 7 | 12 | 14 | 12 | 14 | 8 | 8 |
|---|---|---|---|---|---|---|---|
| − 3 | − 5 | − 4 | − 6 | − 5 | − 8 | − 4 | − 7 |

| 16 | 14 | 7 | 14 | 7 | 7 | 12 | 18 |
|---|---|---|---|---|---|---|---|
| − 8 | − 6 | − 3 | − 7 | − 4 | − 2 | − 5 | − 9 |

## 4

**A** 5009 **B** 9035 **C** 8024 **D** 7406 **E** 9004 **F** 8201

## 5

| 16 | 11 | 14 | 12 | 16 | 13 | 17 | 15 |
|---|---|---|---|---|---|---|---|
| − 9 | − 9 | − 9 | − 9 | − 9 | − 9 | − 9 | − 9 |

## 6

| **A** | **B** | **C** | **D** | **E** |
|---|---|---|---|---|
| 502 | 704 | 304 | 606 | 402 |
| − 395 | − 186 | − 118 | − 408 | − 123 |

## 7

**A**
6 { 2, ☐ ______

**B**
☐ { 5, 3 ______

**C**
7 { 5, ☐ ______

**D**
☐ { 8, 1 ______

**E**
☐ { 6, 4 ______

**F**
8 { 1, ☐ ______

## 8

**A** $353 - 352$

**B** $698 - 594$

**C** $5746 - 5731$

**D** $3826 - 3746$

**E** $7254 - 7048$

**F** $6432 - 1725$

**G** $6289 - 1378$

**H** $408 + 8$

**I** $9068 - 8949$

**J** $30 + 75$

**K** $90 - 45$

**L** $7214 + 86$

**M** $5246 - 638$

**N** $9437 - 7607$

**O** $740 + 259$

## 9

**A** 70 − 1 = ______

**B** 30 − 1 = ______

**C** 80 − 1 = ______

**D** 20 − 1 = ______

**E** 60 − 1 = ______

**F** 90 − 1 = ______

**G** 40 − 1 = ______

**H** 70 − 1 = ______

# Lesson 26

| Facts | + | Problems | + | Bonus | = | TOTAL |
|---|---|---|---|---|---|---|
| | | | | | | |

**1**

| | | | | | | | |
|---|---|---|---|---|---|---|---|
| 18 − 9 | 15 − 9 | 11 − 9 | 17 − 9 | 13 − 9 | 12 − 9 | 16 − 9 | 11 − 9 |

**2**

| | | | | | | | |
|---|---|---|---|---|---|---|---|
| 16 − 7 | 10 − 5 | 16 − 8 | 16 − 9 | 16 − 7 | 10 − 3 | 16 − 10 | 16 − 9 |
| 10 − 4 | 16 − 9 | 10 − 2 | 16 − 7 | 16 − 8 | 10 − 3 | 16 − 9 | 16 − 7 |

**3**

| | | | | | | | |
|---|---|---|---|---|---|---|---|
| 7 − 5 | 7 − 4 | 7 − 1 | 7 − 2 | 7 − 3 | 7 − 6 | 7 − 4 | 7 − 7 |
| 7 − 6 | 7 − 3 | 7 − 0 | 7 − 5 | 7 − 2 | 7 − 4 | 7 − 3 | 7 − 1 |

**4**

| | | | | | | | |
|---|---|---|---|---|---|---|---|
| 12 − 5 | 14 − 6 | 18 − 9 | 12 − 3 | 6 − 1 | 8 − 0 | 12 − 5 | 4 − 4 |
| 12 − 10 | 12 − 4 | 14 − 10 | 12 − 7 | 6 − 5 | 14 − 8 | 18 − 10 | 12 − 8 |

**5**

A 4064 B 5002 C 9400 D 8006 E 6010 F 8000

## 6

A
$$\begin{array}{r} 502 \\ -\ 314 \\ \hline \end{array}$$

B
$$\begin{array}{r} 806 \\ -\ 198 \\ \hline \end{array}$$

C
$$\begin{array}{r} 302 \\ -\ 189 \\ \hline \end{array}$$

D
$$\begin{array}{r} 802 \\ -\ 716 \\ \hline \end{array}$$

E
$$\begin{array}{r} 406 \\ -\ 218 \\ \hline \end{array}$$

## 7

A
$$\square \begin{cases} 4 \\ 3 \end{cases}$$ _______________

B
$$7 \begin{cases} 2 \\ \square \end{cases}$$ _______________

C
$$9 \begin{cases} 8 \\ \square \end{cases}$$ _______________

D
$$\square \begin{cases} 8 \\ 1 \end{cases}$$ _______________

E
$$14 \begin{cases} 8 \\ \square \end{cases}$$ _______________

F
$$\square \begin{cases} 5 \\ 2 \end{cases}$$ _______________

## 8

A
$$\begin{array}{r} 9924 \\ -\ 9852 \\ \hline \end{array}$$

B
$$\begin{array}{r} 7453 \\ -\ 6800 \\ \hline \end{array}$$

C
$$\begin{array}{r} 2532 \\ +\ 2427 \\ \hline \end{array}$$

D
$$\begin{array}{r} 934 \\ -\ 826 \\ \hline \end{array}$$

E
$$\begin{array}{r} 5824 \\ -\ 4904 \\ \hline \end{array}$$

F
$$\begin{array}{r} 3492 \\ -\ \ \ 76 \\ \hline \end{array}$$

G
$$\begin{array}{r} 907 \\ +\ \ 57 \\ \hline \end{array}$$

H
$$\begin{array}{r} 8280 \\ -\ 7219 \\ \hline \end{array}$$

I
$$\begin{array}{r} 480 \\ -\ 421 \\ \hline \end{array}$$

J
$$\begin{array}{r} 6032 \\ +\ \ 540 \\ \hline \end{array}$$

K
$$\begin{array}{r} 4210 \\ +\ \ 606 \\ \hline \end{array}$$

L
$$\begin{array}{r} 3812 \\ -\ 2910 \\ \hline \end{array}$$

M
$$\begin{array}{r} 3846 \\ -\ 3786 \\ \hline \end{array}$$

N
$$\begin{array}{r} 564 \\ -\ 510 \\ \hline \end{array}$$

O
$$\begin{array}{r} 856 \\ +\ \ \ 6 \\ \hline \end{array}$$

# Lesson 27

| Facts | + | Problems | + | Bonus | = | TOTAL |
|---|---|---|---|---|---|---|
| | | | | | | |

## 1

| | | | | | | | |
|---|---|---|---|---|---|---|---|
| 17 − 9 | 15 − 9 | 12 − 9 | 17 − 9 | 13 − 9 | 16 − 9 | 14 − 9 | 18 − 9 |
| 16 − 9 | 13 − 9 | 12 − 9 | 11 − 9 | 15 − 9 | 14 − 9 | 17 − 9 | 12 − 9 |

## 2

| | | | | | | | |
|---|---|---|---|---|---|---|---|
| 8 − 3 | 16 − 7 | 16 − 10 | 8 − 5 | 16 − 9 | 16 − 7 | 8 − 3 | 8 − 5 |
| 16 − 9 | 16 − 7 | 8 − 3 | 8 − 7 | 16 − 7 | 8 − 5 | 8 − 8 | 8 − 3 |

## 3

| | | | | | | | |
|---|---|---|---|---|---|---|---|
| 10 − 4 | 7 − 4 | 12 − 8 | 12 − 3 | 10 − 3 | 7 − 5 | 12 − 7 | 10 − 2 |
| 12 − 4 | 10 − 3 | 7 − 6 | 12 − 8 | 10 − 4 | 10 − 9 | 12 − 4 | 7 − 5 |
| 12 − 3 | 7 − 2 | 12 − 5 | 10 − 3 | 7 − 2 | 10 − 2 | 12 − 3 | 12 − 8 |
| 10 − 5 | 12 − 5 | 7 − 3 | 7 − 7 | 12 − 4 | 12 − 10 | 16 − 8 | 7 − 4 |

## 4

**A**

$$12 \begin{cases} 5 \\ \square \end{cases}$$

**B**

$$7 \begin{cases} 6 \\ \square \end{cases}$$

**C**

$$\square \begin{cases} 9 \\ 1 \end{cases}$$

**D**

$$\square \begin{cases} 7 \\ 3 \end{cases}$$

**E**

$$7 \begin{cases} 2 \\ \square \end{cases}$$

**F**

$$\square \begin{cases} 10 \\ 1 \end{cases}$$

## 5

**A** 4006 **B** 508 **C** 300 **D** 3000 **E** 502 **F** 5020

## 6

**A**
$$\begin{array}{r} 4046 \\ -\ 2185 \\ \hline \end{array}$$

**B**
$$\begin{array}{r} 5024 \\ -\ 1853 \\ \hline \end{array}$$

**C**
$$\begin{array}{r} 3704 \\ -\ 2198 \\ \hline \end{array}$$

**D**
$$\begin{array}{r} 5029 \\ -\ 3641 \\ \hline \end{array}$$

**E**
$$\begin{array}{r} 904 \\ -\ 186 \\ \hline \end{array}$$

## 7

**A**
$$\begin{array}{r} 3460 \\ -\ 2751 \\ \hline \end{array}$$

**B**
$$\begin{array}{r} 6224 \\ -\ 5160 \\ \hline \end{array}$$

**C**
$$\begin{array}{r} 805 \\ +\ 95 \\ \hline \end{array}$$

**D**
$$\begin{array}{r} 420 \\ -\ 311 \\ \hline \end{array}$$

Part 7 continues on the next page.

**E**
$$\begin{array}{r} 948 \\ +\ \ 28 \\ \hline \end{array}$$

**F**
$$\begin{array}{r} 5854 \\ -2917 \\ \hline \end{array}$$

**G**
$$\begin{array}{r} 6953 \\ +\ \ 812 \\ \hline \end{array}$$

**H**
$$\begin{array}{r} 992 \\ -945 \\ \hline \end{array}$$

**I**
$$\begin{array}{r} 8470 \\ -\ \ 815 \\ \hline \end{array}$$

**J**
$$\begin{array}{r} 3480 \\ +1581 \\ \hline \end{array}$$

**K**
$$\begin{array}{r} 5480 \\ -4461 \\ \hline \end{array}$$

**L**
$$\begin{array}{r} 3627 \\ -\ \ \ \ 61 \\ \hline \end{array}$$

**M**
$$\begin{array}{r} 384 \\ +\ \ 26 \\ \hline \end{array}$$

**N**
$$\begin{array}{r} 4826 \\ +\ \ \ \ 26 \\ \hline \end{array}$$

**O**
$$\begin{array}{r} 8352 \\ -4335 \\ \hline \end{array}$$

**P**
$$\begin{array}{r} 9958 \\ -9139 \\ \hline \end{array}$$

**Q**
$$\begin{array}{r} 5260 \\ -3619 \\ \hline \end{array}$$

**R**
$$\begin{array}{r} 8356 \\ +\ \ 156 \\ \hline \end{array}$$

**S**
$$\begin{array}{r} 3860 \\ -\ \ \ \ 45 \\ \hline \end{array}$$

**T**
$$\begin{array}{r} 75 \\ +70 \\ \hline \end{array}$$

# Lesson 28

Facts + Problems + Bonus = TOTAL

## 1

$15 - 9$ $12 - 9$ $16 - 9$ $14 - 9$ $17 - 9$ $13 - 9$ $18 - 9$ $11 - 9$

$16 - 9$ $14 - 9$ $17 - 9$ $13 - 9$ $18 - 9$ $12 - 9$ $15 - 9$ $11 - 9$

## 2

$16 - 7$ $8 - 5$ $8 - 7$ $8 - 3$ $16 - 9$ $16 - 7$ $8 - 3$ $8 - 5$

$16 - 8$ $16 - 7$ $8 - 8$ $8 - 5$ $16 - 9$ $8 - 3$ $8 - 5$ $16 - 7$

## 3

$12 - 8$ $7 - 3$ $10 - 4$ $12 - 4$ $10 - 3$ $12 - 5$ $7 - 5$ $12 - 3$

$14 - 8$ $10 - 2$ $18 - 9$ $14 - 6$ $12 - 7$ $14 - 6$ $10 - 4$ $16 - 8$

$12 - 7$ $7 - 3$ $12 - 4$ $7 - 6$ $12 - 3$ $7 - 4$ $12 - 8$ $10 - 2$

$7 - 5$ $12 - 3$ $10 - 2$ $14 - 8$ $12 - 5$ $8 - 8$ $7 - 7$ $12 - 6$

## 4

**A** 9 { 8, [ ] } ------------------

**B** [ ] { 8, 6 } ------------------

**C** [ ] { 3, 9 } ------------------

**D** 3 { 1, [ ] } ------------------

**E** [ ] { 8, 4 } ------------------

**F** 7 { 5, [ ] } ------------------

## 5

**A** 2045 **B** 2405 **C** 8001 **D** 920 **E** 4054 **F** 1099

## 6

**A**
```
  3045
−  165
```

**B**
```
  4029
− 2850
```

**C**
```
  4206
− 3198
```

**D**
```
  502
− 115
```

**E**
```
  3506
− 1208
```

## 7

**A**
```
  3824
− 2917
```

**B**
```
  6520
− 5350
```

**C**
```
  5350
− 4324
```

**D**
```
  1956
−   18
```

Part 7 continues on the next page.

**E**
$$\begin{array}{r} 5490 \\ -\,4472 \\ \hline \end{array}$$

**F**
$$\begin{array}{r} 4615 \\ +\quad 605 \\ \hline \end{array}$$

**G**
$$\begin{array}{r} 3265 \\ +\quad 15 \\ \hline \end{array}$$

**H**
$$\begin{array}{r} 482 \\ -\,474 \\ \hline \end{array}$$

**I**
$$\begin{array}{r} 9432 \\ -\,8726 \\ \hline \end{array}$$

**J**
$$\begin{array}{r} 384 \\ -\,306 \\ \hline \end{array}$$

**K**
$$\begin{array}{r} 5648 \\ -\quad 839 \\ \hline \end{array}$$

**L**
$$\begin{array}{r} 482 \\ -\,382 \\ \hline \end{array}$$

**M**
$$\begin{array}{r} 374 \\ +\,186 \\ \hline \end{array}$$

**N**
$$\begin{array}{r} 94 \\ +\,80 \\ \hline \end{array}$$

**O**
$$\begin{array}{r} 702 \\ +\,288 \\ \hline \end{array}$$

**P**
$$\begin{array}{r} 7410 \\ +\quad 609 \\ \hline \end{array}$$

**Q**
$$\begin{array}{r} 396 \\ -\quad 8 \\ \hline \end{array}$$

**R**
$$\begin{array}{r} 6842 \\ -\quad 71 \\ \hline \end{array}$$

**S**
$$\begin{array}{r} 463 \\ -\quad 80 \\ \hline \end{array}$$

**T**
$$\begin{array}{r} 4588 \\ +\quad 28 \\ \hline \end{array}$$

# Lesson 29

| Facts | + | Problems | + | Bonus | = | TOTAL |
|---|---|---|---|---|---|---|
| | | | | | | |

## 1

| | | | | | | | |
|---|---|---|---|---|---|---|---|
| 15 − 9 | 13 − 9 | 11 − 9 | 18 − 9 | 16 − 9 | 15 − 9 | 14 − 9 | 12 − 9 |
| 18 − 9 | 11 − 9 | 14 − 9 | 17 − 9 | 15 − 9 | 12 − 9 | 16 − 9 | 13 − 9 |

## 2

| | | | | | | | |
|---|---|---|---|---|---|---|---|
| 12 − 7 | 10 − 2 | 10 − 4 | 12 − 8 | 7 − 4 | 12 − 9 | 10 − 3 | 7 − 5 |
| 10 − 2 | 12 − 3 | 10 − 3 | 12 − 7 | 7 − 4 | 12 − 3 | 10 − 4 | 12 − 6 |
| 12 − 5 | 7 − 5 | 10 − 4 | 14 − 8 | 12 − 4 | 14 − 6 | 12 − 8 | 12 − 7 |
| 14 − 6 | 12 − 5 | 18 − 9 | 10 − 3 | 12 − 4 | 14 − 8 | 14 − 7 | 12 − 8 |

## 3

| | | | | | | | |
|---|---|---|---|---|---|---|---|
| 16 − 7 | 16 − 8 | 8 − 3 | 16 − 9 | 8 − 5 | 16 − 7 | 8 − 3 | 16 − 9 |

## 4

**A** 4036 **B** 8214 **C** 7040 **D** 3008 **E** 6208

## 5

**A**

The big number is 9. A small number is 4.

**B**

A small number is 10. Another small number is 4.

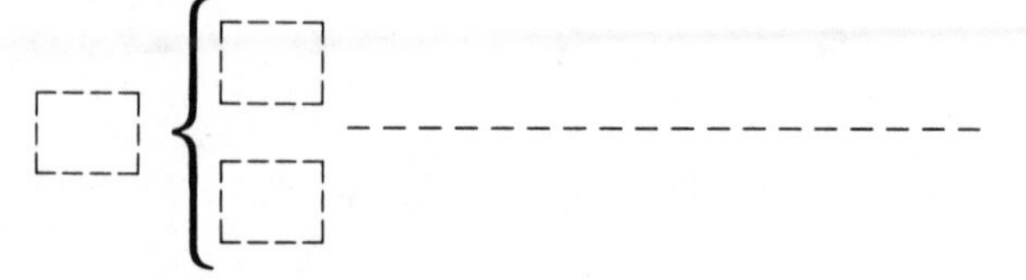

**C**

The big number is 5. A small number is 3.

**D**

A small number is 8. Another small number is 2.

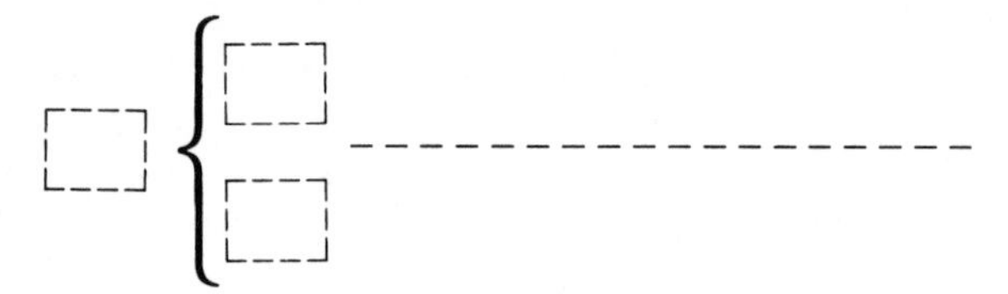

**E**

A small number is 9. Another small number is 1.

**F**

The big number is 7. A small number is 3.

## 6

**A**
$$\begin{array}{r} 502 \\ -\ 185 \\ \hline \end{array}$$

**B**
$$\begin{array}{r} 308 \\ -\ 157 \\ \hline \end{array}$$

**C**
$$\begin{array}{r} 5049 \\ -\ 768 \\ \hline \end{array}$$

**D**
$$\begin{array}{r} 7082 \\ -\ 3140 \\ \hline \end{array}$$

**E**
$$\begin{array}{r} 3068 \\ -\ 184 \\ \hline \end{array}$$

## 7

**A**
$$\begin{array}{r} 7632 \\ -\ 808 \\ \hline \end{array}$$

**B**
$$\begin{array}{r} 9428 \\ -\ 50 \\ \hline \end{array}$$

**C**
$$\begin{array}{r} 825 \\ +\ 75 \\ \hline \end{array}$$

**D**
$$\begin{array}{r} 3420 \\ -\ 619 \\ \hline \end{array}$$

**E**
$$\begin{array}{r} 8425 \\ +\ 67 \\ \hline \end{array}$$

**F**
$$\begin{array}{r} 70 \\ -45 \\ \hline \end{array}$$

**G**
$$\begin{array}{r} 850 \\ -710 \\ \hline \end{array}$$

**H**
$$\begin{array}{r} 340 \\ +\ 60 \\ \hline \end{array}$$

**I**
$$\begin{array}{r} 8250 \\ +\ 50 \\ \hline \end{array}$$

**J**
$$\begin{array}{r} 4215 \\ -3710 \\ \hline \end{array}$$

**K**
$$\begin{array}{r} 3040 \\ -\ 931 \\ \hline \end{array}$$

**L**
$$\begin{array}{r} 7410 \\ +\ 609 \\ \hline \end{array}$$

**M**
$$\begin{array}{r} 396 \\ -\ 8 \\ \hline \end{array}$$

**N**
$$\begin{array}{r} 6842 \\ -\ 71 \\ \hline \end{array}$$

**O**
$$\begin{array}{r} 463 \\ -\ 80 \\ \hline \end{array}$$

**P**
$$\begin{array}{r} 4588 \\ +\ 28 \\ \hline \end{array}$$

**Q**
$$\begin{array}{r} 348 \\ -140 \\ \hline \end{array}$$

**R**
$$\begin{array}{r} 9480 \\ -1776 \\ \hline \end{array}$$

**S**
$$\begin{array}{r} 3528 \\ -\ 360 \\ \hline \end{array}$$

**T**
$$\begin{array}{r} 780 \\ +\ 80 \\ \hline \end{array}$$

# Lesson 30

| Test | + | Facts | + | Problems | + | Bonus | = | TOTAL |
|---|---|---|---|---|---|---|---|---|
| | | | | | | | | |

**1**

$\begin{array}{r} 12 \\ -\ 9 \\ \hline \end{array}$ $\begin{array}{r} 17 \\ -\ 9 \\ \hline \end{array}$ $\begin{array}{r} 14 \\ -\ 9 \\ \hline \end{array}$ $\begin{array}{r} 16 \\ -\ 9 \\ \hline \end{array}$ $\begin{array}{r} 11 \\ -\ 9 \\ \hline \end{array}$ $\begin{array}{r} 18 \\ -\ 9 \\ \hline \end{array}$ $\begin{array}{r} 15 \\ -\ 9 \\ \hline \end{array}$ $\begin{array}{r} 13 \\ -\ 9 \\ \hline \end{array}$

$\begin{array}{r} 11 \\ -\ 9 \\ \hline \end{array}$ $\begin{array}{r} 17 \\ -\ 9 \\ \hline \end{array}$ $\begin{array}{r} 13 \\ -\ 9 \\ \hline \end{array}$ $\begin{array}{r} 16 \\ -\ 9 \\ \hline \end{array}$ $\begin{array}{r} 12 \\ -\ 9 \\ \hline \end{array}$ $\begin{array}{r} 15 \\ -\ 9 \\ \hline \end{array}$ $\begin{array}{r} 18 \\ -\ 9 \\ \hline \end{array}$ $\begin{array}{r} 14 \\ -\ 9 \\ \hline \end{array}$

**2**

**A**

The big number is 7. A small number is 4.

**B**

A small number is 10. Another small number is 5.

**C**

A small number is 7. Another small number is 3.

**D**

The big number is 10. A small number is 2.

**E**

The big number is 5. A small number is 4.

**F**

A small number is 8. Another small number is 2.

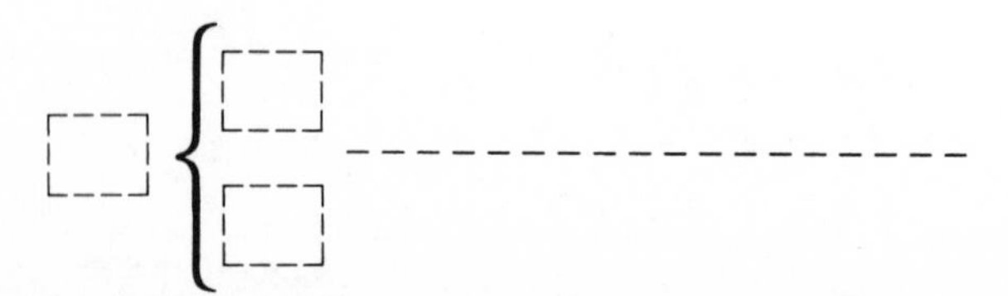

## 3

$12 - 8$ $\quad 7 - 3$ $\quad 12 - 7$ $\quad 12 - 4$ $\quad 12 - 5$ $\quad 7 - 5$ $\quad 10 - 3$ $\quad 14 - 6$

$12 - 3$ $\quad 12 - 7$ $\quad 12 - 5$ $\quad 7 - 6$ $\quad 12 - 8$ $\quad 12 - 4$ $\quad 12 - 5$ $\quad 7 - 5$

## 4

**A** [ ] { 4, 3 } ------------------

**B** 7 { 2, [ ] } ------------------

**C** 9 { 8, [ ] } ------------------

**D** [ ] { 8, 1 } ------------------

**E** 14 { 8, [ ] } ------------------

**F** [ ] { 5, 2 } ------------------

## 5

$16 - 7$ $\quad 8 - 5$ $\quad 16 - 9$ $\quad 8 - 3$ $\quad 16 - 7$ $\quad 8 - 3$ $\quad 16 - 7$ $\quad 8 - 5$

## 6

**A** 1035

**B** 2406

**C** 3006

**D** 5071

**E** 3401

## 7

**A** Ann had 9 oranges. She gave 4 oranges to her friends.

9 is a _______________ number.

4 is a _______________ number.

**B** Jack has 7 books. He buys 2 books.

7 is a _______________ number.

2 is a _______________ number.

**C** Gloria has 4 pens. She gives away 1 pen.

4 is a _______________ number.

1 is a _______________ number.

**D** 6 children are in the park. 4 children go home.

6 is a _______________ number.

4 is a _______________ number.

**E** Roy ate 8 strawberries. Then he ate 5 more.

8 is a _______________ number.

5 is a _______________ number.

**F** Jane has 5 cats. 3 cats run away.

5 is a _______________ number.

3 is a _______________ number.

## 8

**A** $\begin{array}{r} 802 \\ -\ 184 \\ \hline \end{array}$

**B** $\begin{array}{r} 704 \\ -\ 553 \\ \hline \end{array}$

**C** $\begin{array}{r} 6084 \\ -\ 4191 \\ \hline \end{array}$

**D** $\begin{array}{r} 8063 \\ -\ 6153 \\ \hline \end{array}$

**E** $\begin{array}{r} 306 \\ -\ 48 \\ \hline \end{array}$

## 9

**A** $\begin{array}{r} 7496 \\ -\ 1895 \\ \hline \end{array}$

**B** $\begin{array}{r} 4825 \\ +\ 1362 \\ \hline \end{array}$

**C** $\begin{array}{r} 90 \\ +\ 45 \\ \hline \end{array}$

**D** $\begin{array}{r} 70 \\ -\ 45 \\ \hline \end{array}$

**E** $\begin{array}{r} 420 \\ -\ 402 \\ \hline \end{array}$

**F** $\begin{array}{r} 4826 \\ +\ 850 \\ \hline \end{array}$

**G** $\begin{array}{r} 7428 \\ -\ 4357 \\ \hline \end{array}$

**H** $\begin{array}{r} 3245 \\ -\ 724 \\ \hline \end{array}$

**I** $\begin{array}{r} 528 \\ +\ 418 \\ \hline \end{array}$

**J** $\begin{array}{r} 9638 \\ -\ 829 \\ \hline \end{array}$

# Lesson 31

Facts + Problems + Bonus = TOTAL

## 1

| A | B | C | D | E |
|---|---|---|---|---|
| 6047 | 9420 | 8006 | 4090 | 8201 |

## 2

$$\begin{array}{r}14\\-\ 9\\\hline\end{array}\quad\begin{array}{r}14\\-\ 7\\\hline\end{array}\quad\begin{array}{r}12\\-\ 9\\\hline\end{array}\quad\begin{array}{r}12\\-10\\\hline\end{array}\quad\begin{array}{r}16\\-\ 9\\\hline\end{array}\quad\begin{array}{r}18\\-\ 9\\\hline\end{array}\quad\begin{array}{r}18\\-10\\\hline\end{array}\quad\begin{array}{r}15\\-\ 9\\\hline\end{array}$$

$$\begin{array}{r}11\\-\ 9\\\hline\end{array}\quad\begin{array}{r}13\\-\ 9\\\hline\end{array}\quad\begin{array}{r}16\\-\ 9\\\hline\end{array}\quad\begin{array}{r}16\\-\ 8\\\hline\end{array}\quad\begin{array}{r}15\\-10\\\hline\end{array}\quad\begin{array}{r}15\\-\ 9\\\hline\end{array}\quad\begin{array}{r}17\\-\ 9\\\hline\end{array}\quad\begin{array}{r}11\\-\ 9\\\hline\end{array}$$

## 3

**A** 10 { 2 ________ ; ☐ ________

**B** 10 { 4 ________ ; ☐ ________

**C** 10 { 3 ________ ; ☐ ________

**D** 10 { 1 ________ ; ☐ ________

## 4

$$\begin{array}{r}10\\-\ 7\\\hline\end{array}\quad\begin{array}{r}10\\-\ 6\\\hline\end{array}\quad\begin{array}{r}10\\-\ 9\\\hline\end{array}\quad\begin{array}{r}10\\-\ 8\\\hline\end{array}\quad\begin{array}{r}10\\-\ 6\\\hline\end{array}\quad\begin{array}{r}10\\-\ 9\\\hline\end{array}\quad\begin{array}{r}10\\-\ 7\\\hline\end{array}\quad\begin{array}{r}10\\-\ 8\\\hline\end{array}$$

$$\begin{array}{r}10\\-\ 3\\\hline\end{array}\quad\begin{array}{r}10\\-\ 7\\\hline\end{array}\quad\begin{array}{r}10\\-\ 2\\\hline\end{array}\quad\begin{array}{r}10\\-\ 4\\\hline\end{array}\quad\begin{array}{r}10\\-\ 8\\\hline\end{array}\quad\begin{array}{r}10\\-\ 6\\\hline\end{array}\quad\begin{array}{r}10\\-\ 9\\\hline\end{array}\quad\begin{array}{r}10\\-\ 1\\\hline\end{array}$$

## 5

| | | | | | | | |
|---|---|---|---|---|---|---|---|
| $16 - 7$ | $8 - 3$ | $12 - 7$ | $12 - 8$ | $16 - 9$ | $7 - 4$ | $8 - 5$ | $10 - 2$ |
| $12 - 4$ | $10 - 2$ | $12 - 8$ | $12 - 7$ | $16 - 7$ | $12 - 3$ | $10 - 3$ | $8 - 5$ |
| $16 - 7$ | $10 - 4$ | $12 - 9$ | $12 - 6$ | $14 - 8$ | $8 - 3$ | $14 - 6$ | $12 - 3$ |
| $12 - 7$ | $14 - 8$ | $10 - 4$ | $14 - 6$ | $12 - 4$ | $12 - 9$ | $12 - 5$ | $12 - 8$ |

## 6

**A**
Jill had 7 frogs. She lost 3 frogs.

7 is a ________________ number.

3 is a ________________ number.

**B**
Mr. Deloria had 7 books. He bought 3 books.

7 is a ________________ number.

3 is a ________________ number.

**C**
Tim had 6 pencils. He found 3 pencils.

6 is a ________________ number.

3 is a ________________ number.

**D**
Gino made 5 sandwiches. He gave away 4 sandwiches.

5 is a ________________ number.

4 is a ________________ number.

## 7

**A**
Al found 8 strawberries in the refrigerator. He ate 3 strawberries. How many strawberries are in the refrigerator now?

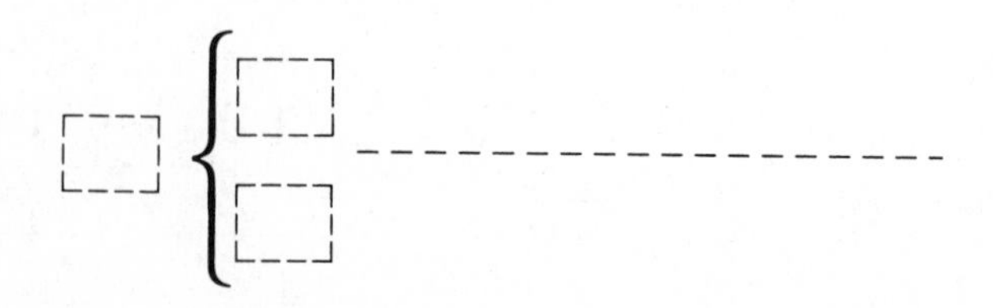

Part 7 continues on the next page.

**B**
Ann has 5 magazines. She buys 4 more magazines. How many magazines does she have now?

**C**
Bill had 6 tops. Then he got 2 tops for his birthday. How many tops did Bill have altogether?

**D**
Tom had 9 chocolates. He gave 3 chocolates to a friend. How many chocolates does he have left?

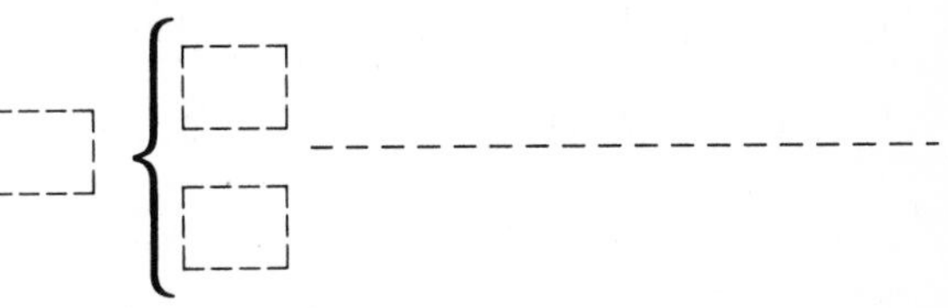

**E**
Gloria had 7 apples. She ate 3 apples. How many apples does she have left?

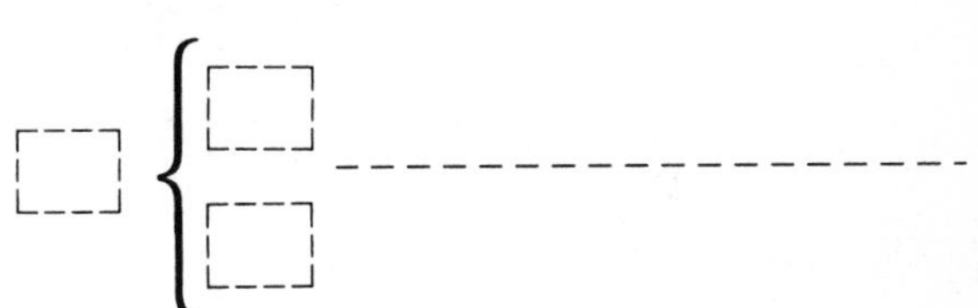

## 8

**A** $\begin{array}{r} 504 \\ -\ 130 \\ \hline \end{array}$

**B** $\begin{array}{r} 702 \\ -\ 585 \\ \hline \end{array}$

**C** $\begin{array}{r} 4064 \\ -\ 1284 \\ \hline \end{array}$

**D** $\begin{array}{r} 9067 \\ -\ 7361 \\ \hline \end{array}$

**E** $\begin{array}{r} 3504 \\ -\ \ 328 \\ \hline \end{array}$

## 9

**A** $\begin{array}{r} 4230 \\ -\ 3823 \\ \hline \end{array}$

**B** $\begin{array}{r} 4826 \\ +\ \ 177 \\ \hline \end{array}$

**C** $\begin{array}{r} 4427 \\ -\ \ 624 \\ \hline \end{array}$

**D** $\begin{array}{r} 50 \\ -\ 15 \\ \hline \end{array}$

**E** $\begin{array}{r} 785 \\ -\ \ 35 \\ \hline \end{array}$

**F** $\begin{array}{r} 7265 \\ -\ 6180 \\ \hline \end{array}$

**G** $\begin{array}{r} 384 \\ -\ 376 \\ \hline \end{array}$

**H** $\begin{array}{r} 4280 \\ +\ 1786 \\ \hline \end{array}$

**I** $\begin{array}{r} 280 \\ +\ 1280 \\ \hline \end{array}$

**J** $\begin{array}{r} 946 \\ -\ 870 \\ \hline \end{array}$

# Lesson 32

| Test | + | Facts | + | Problems | + | Bonus | = | TOTAL |
|---|---|---|---|---|---|---|---|---|
| | | | | | | | | |

## 1

| 5 | 5 | 5 | 5 | 5 | 5 | 5 | 5 |
|---|---|---|---|---|---|---|---|
| − 3 | − 1 | − 2 | − 5 | − 2 | − 4 | − 3 | − 1 |

## 2

| | | | | |
|---|---|---|---|---|
| A | 5001 | 4064 | 2505 | 340 |
| B | 9052 | 7008 | 520 | 3706 |
| C | 980 | 4506 | 7009 | 4031 |
| D | 6102 | 870 | 9003 | 6041 |
| E | 750 | 8002 | 5013 | 4106 |
| F | 8309 | 260 | 5004 | 3026 |

## 3

| 14 | 14 | 13 | 11 | 16 | 12 | 12 | 15 |
|---|---|---|---|---|---|---|---|
| − 9 | − 7 | − 9 | − 9 | − 8 | − 9 | − 5 | − 9 |

| 18 | 17 | 17 | 10 | 13 | 16 | 12 | 15 |
|---|---|---|---|---|---|---|---|
| − 9 | − 10 | − 9 | − 9 | − 9 | − 8 | − 9 | − 9 |

## 4

**A** 10 { 3 ______ ; [ ] ______

**B** 10 { 4 ______ ; [ ] ______

**C** 10 { 2 ______ ; [ ] ______

**D** 10 { 5 ______ ; [ ] ______

## 5

| | | | | | | | |
|---|---|---|---|---|---|---|---|
| $10 - 8$ | $10 - 6$ | $10 - 5$ | $10 - 7$ | $10 - 6$ | $10 - 9$ | $10 - 8$ | $10 - 6$ |
| $10 - 2$ | $10 - 4$ | $10 - 7$ | $10 - 3$ | $10 - 6$ | $10 - 8$ | $10 - 1$ | $10 - 9$ |

## 6

| | | | | | | | |
|---|---|---|---|---|---|---|---|
| $16 - 7$ | $12 - 8$ | $10 - 3$ | $8 - 3$ | $12 - 4$ | $12 - 7$ | $8 - 5$ | $10 - 4$ |
| $7 - 2$ | $12 - 6$ | $12 - 4$ | $16 - 7$ | $7 - 4$ | $12 - 8$ | $8 - 5$ | $12 - 4$ |
| $14 - 6$ | $16 - 7$ | $12 - 7$ | $8 - 3$ | $7 - 5$ | $7 - 7$ | $12 - 5$ | $7 - 3$ |

## 7

**A**
Jane had 9 marbles. Her brother gave her 3 more marbles. How many marbles did she have in all?

**B**
Sam bought 7 shirts. He lost 3 shirts. How many shirts does he have left?

**C**
Joan and Ann baked 4 pies on Friday. On Saturday they baked 5 more pies. How many pies did they bake?

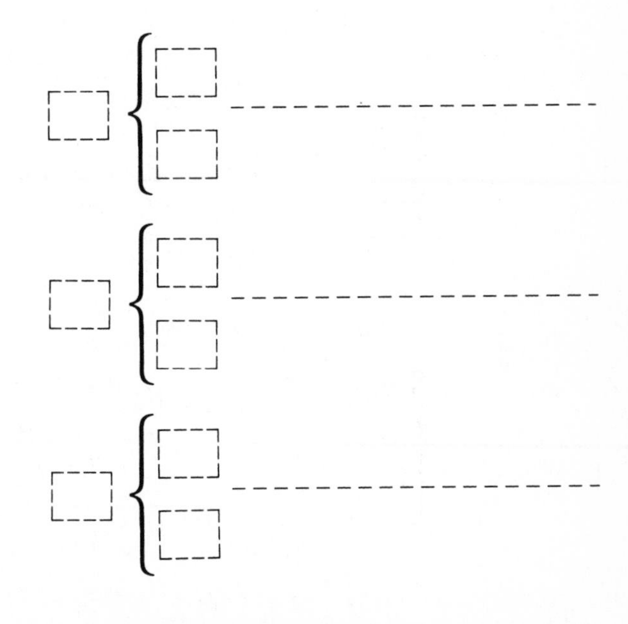

Part 7 continues on the next page.

**D**

Bob helped his mother make 7 hamburgers. Then he ate 3 of them. How many hamburgers were left?

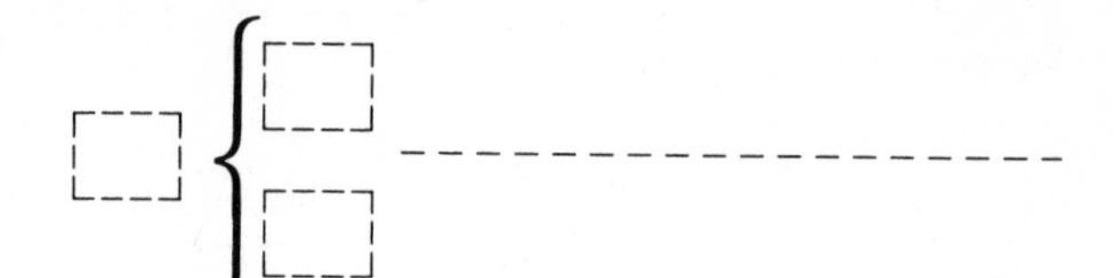

**E**

You have 8 seashells. A friend gives you 3 more. How many seashells do you have in all?

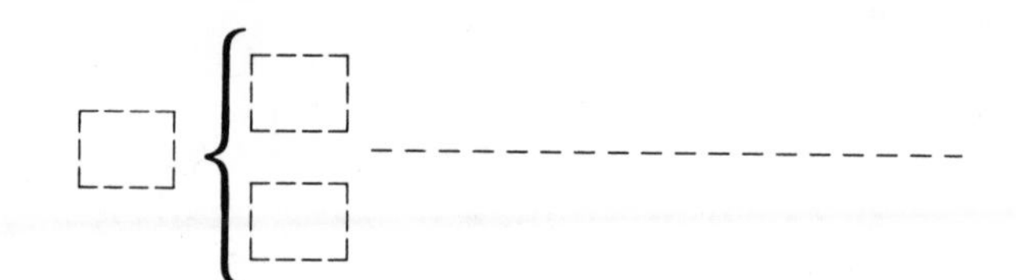

## 8

**A** $\begin{array}{r} 532 \\ -\ 239 \\ \hline \end{array}$

**B** $\begin{array}{r} 512 \\ -\ \ 18 \\ \hline \end{array}$

**C** $\begin{array}{r} 8926 \\ -\ \ 960 \\ \hline \end{array}$

**D** $\begin{array}{r} 914 \\ -\ 146 \\ \hline \end{array}$

**E** $\begin{array}{r} 5550 \\ -\ 5171 \\ \hline \end{array}$

## 9

**A** $\begin{array}{r} 4026 \\ -\ 3156 \\ \hline \end{array}$

**B** $\begin{array}{r} 3846 \\ +\ \ 256 \\ \hline \end{array}$

**C** $\begin{array}{r} 8038 \\ -\ \ 527 \\ \hline \end{array}$

**D** $\begin{array}{r} 406 \\ -\ 398 \\ \hline \end{array}$

**E** $\begin{array}{r} 705 \\ -\ \ 10 \\ \hline \end{array}$

**F** $\begin{array}{r} 8042 \\ +\ \ 160 \\ \hline \end{array}$

**G** $\begin{array}{r} 9280 \\ -\ 8665 \\ \hline \end{array}$

**H** $\begin{array}{r} 602 \\ -\ \ 92 \\ \hline \end{array}$

**I** $\begin{array}{r} 590 \\ -\ 186 \\ \hline \end{array}$

**J** $\begin{array}{r} 6804 \\ -\ 1698 \\ \hline \end{array}$

**K** $\begin{array}{r} 40 \\ -\ 13 \\ \hline \end{array}$

**L** $\begin{array}{r} 560 \\ -\ 480 \\ \hline \end{array}$

**M** $\begin{array}{r} 108 \\ +\ \ 18 \\ \hline \end{array}$

**N** $\begin{array}{r} 4026 \\ -\ 3908 \\ \hline \end{array}$

**O** $\begin{array}{r} 402 \\ -\ \ \ 6 \\ \hline \end{array}$

**P** $\begin{array}{r} 9437 \\ -\ 7607 \\ \hline \end{array}$

# Lesson 33

| Test | + | Facts | + | Problems | + | Bonus | = | TOTAL |
|---|---|---|---|---|---|---|---|---|
| | | | | | | | | |

## 1

| 13 | 13 | 13 | 13 | 13 | 13 | 13 | 13 |
|---|---|---|---|---|---|---|---|
| − 5 | − 10 | − 8 | − 9 | − 5 | − 8 | − 10 | − 5 |

## 2

**A** 10 { 4 ______ / ☐ ______

**B** 10 { 2 ______ / ☐ ______

**C** 10 { 3 ______ / ☐ ______

**D** 10 { 1 ______ / ☐ ______

## 3

| 10 | 10 | 10 | 10 | 10 | 10 | 10 | 10 |
|---|---|---|---|---|---|---|---|
| − 6 | − 2 | − 4 | − 8 | − 3 | − 7 | − 5 | − 9 |

| 10 | 10 | 10 | 10 | 10 | 10 | 10 | 10 |
|---|---|---|---|---|---|---|---|
| − 4 | − 2 | − 7 | − 5 | − 8 | − 3 | − 6 | − 4 |

## 4

| 16 | 15 | 8 | 14 | 12 | 16 | 12 | 11 |
|---|---|---|---|---|---|---|---|
| − 7 | − 9 | − 3 | − 9 | − 7 | − 9 | − 8 | − 9 |

| 12 | 14 | 16 | 12 | 17 | 12 | 8 | 15 |
|---|---|---|---|---|---|---|---|
| − 7 | − 6 | − 7 | − 3 | − 9 | − 8 | − 3 | − 9 |

Part 4 continues on the next page.

$\begin{array}{r} 13 \\ -\ 9 \\ \hline \end{array}$ $\begin{array}{r} 12 \\ -\ 6 \\ \hline \end{array}$ $\begin{array}{r} 8 \\ -5 \\ \hline \end{array}$ $\begin{array}{r} 10 \\ -\ 4 \\ \hline \end{array}$ $\begin{array}{r} 12 \\ -\ 4 \\ \hline \end{array}$ $\begin{array}{r} 14 \\ -\ 8 \\ \hline \end{array}$ $\begin{array}{r} 12 \\ -\ 9 \\ \hline \end{array}$ $\begin{array}{r} 8 \\ -3 \\ \hline \end{array}$

$\begin{array}{r} 15 \\ -\ 9 \\ \hline \end{array}$ $\begin{array}{r} 10 \\ -\ 3 \\ \hline \end{array}$ $\begin{array}{r} 12 \\ -\ 5 \\ \hline \end{array}$ $\begin{array}{r} 16 \\ -\ 7 \\ \hline \end{array}$ $\begin{array}{r} 8 \\ -5 \\ \hline \end{array}$ $\begin{array}{r} 10 \\ -\ 2 \\ \hline \end{array}$ $\begin{array}{r} 14 \\ -\ 7 \\ \hline \end{array}$ $\begin{array}{r} 13 \\ -\ 9 \\ \hline \end{array}$

## 5

$\begin{array}{r} 5 \\ -2 \\ \hline \end{array}$ $\begin{array}{r} 5 \\ -5 \\ \hline \end{array}$ $\begin{array}{r} 5 \\ -3 \\ \hline \end{array}$ $\begin{array}{r} 5 \\ -0 \\ \hline \end{array}$ $\begin{array}{r} 5 \\ -4 \\ \hline \end{array}$ $\begin{array}{r} 5 \\ -2 \\ \hline \end{array}$ $\begin{array}{r} 5 \\ -1 \\ \hline \end{array}$ $\begin{array}{r} 5 \\ -3 \\ \hline \end{array}$

## 6

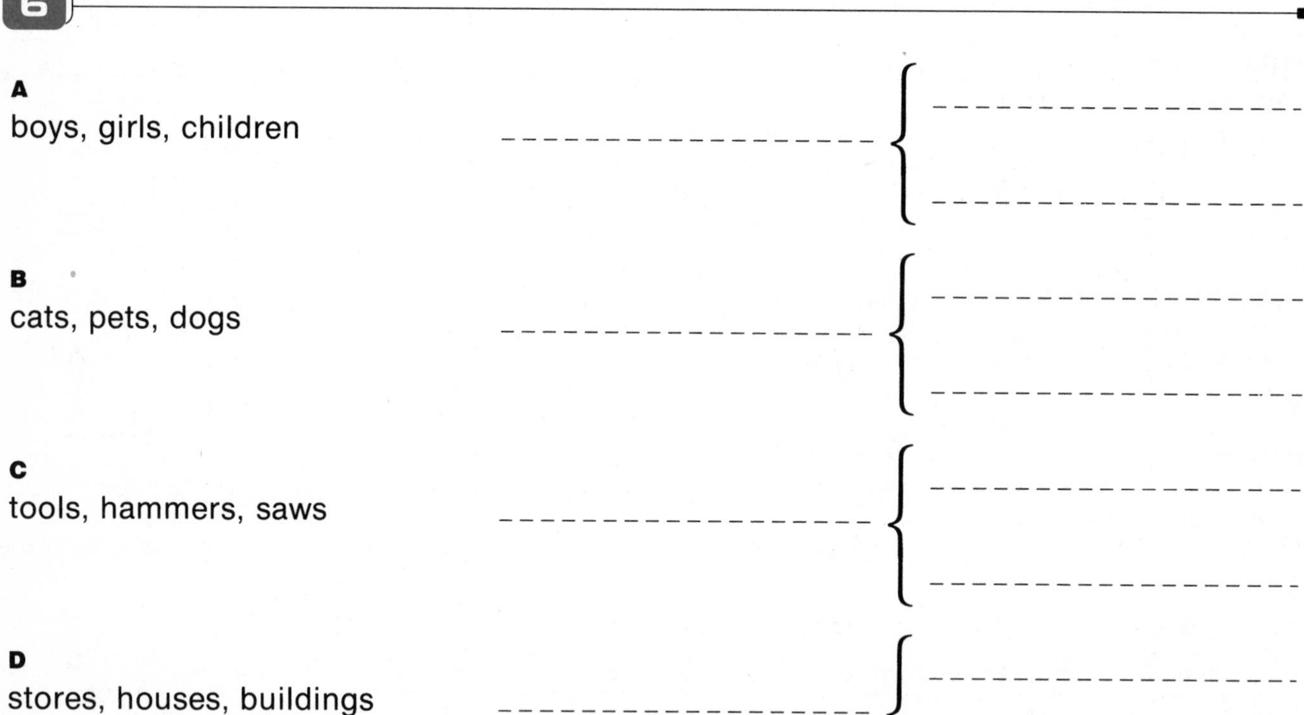

**A**
boys, girls, children ________ { ________ / ________

**B**
cats, pets, dogs ________ { ________ / ________

**C**
tools, hammers, saws ________ { ________ / ________

**D**
stores, houses, buildings ________ { ________ / ________

**E**
shirts, clothes, dresses ________ { ________ / ________

## 7

**A**

$$\begin{array}{r} 630 \\ -\ \ 35 \\ \hline \end{array}$$

**B**

$$\begin{array}{r} 4926 \\ -3749 \\ \hline \end{array}$$

**C**

$$\begin{array}{r} 712 \\ -665 \\ \hline \end{array}$$

**D**

$$\begin{array}{r} 6810 \\ -5119 \\ \hline \end{array}$$

**E**

$$\begin{array}{r} 474 \\ -\ \ 87 \\ \hline \end{array}$$

## 8

**A**

Stefan had 6 card games. He bought 2 more. How many card games did Stefan have in all?

**B**

Lisa and a friend are putting up a fence. The fence has 4 gates. Lisa and her friend have finished 3 gates. How many gates do they have left?

**C**

Su Lin wrote a story about 12 giraffes. Then Su Lin changed the story. She took out 3 giraffes. How many giraffes did she have left in her story?

**D**

Mrs. Tibbs, who plays golf, had 9 golf balls. She got 4 more for her birthday. How many golf balls does she have now?

**E**

There are 12 girls playing tennis. 4 go home. How many girls are left?

**F**

Our cats had 9 kittens. We gave 2 kittens to a friend. How many kittens did we have left?

**G**

Mrs. Rojas wrote 5 children's stories last year. This year she has written 7. How many stories has Mrs. Rojas written in all?

9

A
$$\begin{array}{r} 4028 \\ -\ 3857 \\ \hline \end{array}$$

B
$$\begin{array}{r} 5084 \\ -\ 2426 \\ \hline \end{array}$$

C
$$\begin{array}{r} 3460 \\ -\ 730 \\ \hline \end{array}$$

D
$$\begin{array}{r} 3820 \\ +\ 188 \\ \hline \end{array}$$

E
$$\begin{array}{r} 508 \\ -\ 27 \\ \hline \end{array}$$

F
$$\begin{array}{r} 3416 \\ +\ 810 \\ \hline \end{array}$$

G
$$\begin{array}{r} 5308 \\ -\ 1248 \\ \hline \end{array}$$

H
$$\begin{array}{r} 6208 \\ -\ 5189 \\ \hline \end{array}$$

I
$$\begin{array}{r} 3040 \\ -\ 931 \\ \hline \end{array}$$

J
$$\begin{array}{r} 7410 \\ +\ 609 \\ \hline \end{array}$$

K
$$\begin{array}{r} 396 \\ -\ 8 \\ \hline \end{array}$$

L
$$\begin{array}{r} 6842 \\ -\ 71 \\ \hline \end{array}$$

M
$$\begin{array}{r} 4588 \\ +\ 28 \\ \hline \end{array}$$

N
$$\begin{array}{r} 348 \\ -\ 140 \\ \hline \end{array}$$

O
$$\begin{array}{r} 9480 \\ -\ 1776 \\ \hline \end{array}$$

P
$$\begin{array}{r} 3528 \\ -\ 360 \\ \hline \end{array}$$

# Lesson 34

| Facts | + | Problems | + | Bonus | = | TOTAL |
|---|---|---|---|---|---|---|
| | | | | | | |

## 1

| | | | | | | | |
|---|---|---|---|---|---|---|---|
| 9 | 9 | 9 | 9 | 9 | 9 | 9 | 9 |
| − 4 | − 1 | − 3 | − 2 | − 4 | − 3 | − 1 | − 2 |

## 2

| | | | | | | | |
|---|---|---|---|---|---|---|---|
| 13 | 5 | 13 | 5 | 5 | 13 | 5 | 13 |
| − 5 | − 2 | − 8 | − 3 | − 2 | − 5 | − 3 | − 8 |

## 3

| | | | | | | | |
|---|---|---|---|---|---|---|---|
| 10 | 10 | 10 | 10 | 10 | 10 | 10 | 10 |
| − 7 | − 8 | − 4 | − 2 | − 5 | − 6 | − 3 | − 1 |

| | | | | | | | |
|---|---|---|---|---|---|---|---|
| 10 | 10 | 10 | 10 | 10 | 10 | 10 | 10 |
| − 6 | − 9 | − 8 | − 3 | − 4 | − 7 | − 2 | − 3 |

## 4

| | | | | | | | |
|---|---|---|---|---|---|---|---|
| 12 | 16 | 12 | 14 | 12 | 12 | 15 | 11 |
| − 9 | − 7 | − 8 | − 9 | − 7 | − 6 | − 9 | − 9 |

| | | | | | | | |
|---|---|---|---|---|---|---|---|
| 7 | 12 | 16 | 12 | 16 | 12 | 13 | 12 |
| − 3 | − 8 | − 9 | − 5 | − 7 | − 3 | − 9 | − 4 |

| | | | | | | | |
|---|---|---|---|---|---|---|---|
| 8 | 12 | 12 | 16 | 7 | 14 | 18 | 12 |
| − 5 | − 9 | − 4 | − 7 | − 4 | − 6 | − 9 | − 3 |

| | | | | | | | |
|---|---|---|---|---|---|---|---|
| 7 | 12 | 16 | 14 | 11 | 8 | 12 | 8 |
| − 5 | − 9 | − 8 | − 8 | − 9 | − 7 | − 4 | − 3 |

## 5

**A**
little boys, big boys, boys

**B**
animals, cows, horses

**C**
students, boys, girls

**D**
red pens, blue pens, pens

**E**
toys, balls, blocks

**F**
clean cups, cups, dirty cups

## 6

**A**
Frank wanted to make a big apple pie. He had 6 apples, but they weren't enough. He bought 2 more. How many apples in all did Frank use?

**B**
Mrs. Ives is an artist. She made 4 paintings last month. She sold 3 of them. How many paintings did she have left?

Part 6 continues on the next page.

**C**
Bob blew up 12 balloons. 3 broke. How many balloons did he have left?

**D**
A woman made 9 baskets last week. This week she finished 4 more. How many baskets has she finished so far?

**E**
Mr. Yakamura had 12 students in his dance class. 8 students went home. How many students were left?

**F**
There were 9 alligators in the river. 2 got out of the river. How many alligators were left in the river?

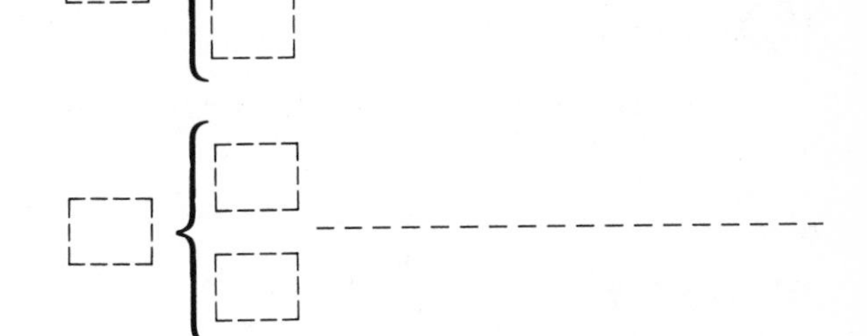

**G**
Marina rode her bike 5 times to a friend's house on Friday. On Saturday she rode 7 times to her friend's house. How many times did Marina ride to her friend's house?

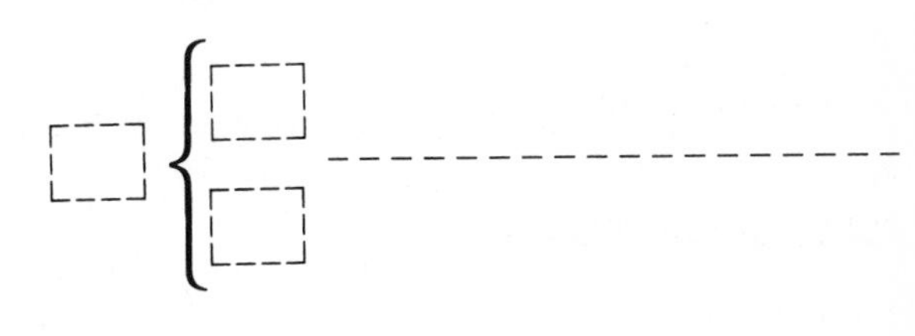

## 7

| A | B | C | D | E |
|---|---|---|---|---|
| 8365 − 1480 | 8080 − 3125 | 4280 + 1786 | 832 + 148 | 8140 − 7470 |

| F | G | H | I | J |
|---|---|---|---|---|
| 9038 − 8969 | 6024 − 750 | 803 − 40 | 7910 − 788 | 8050 − 7868 |

| K | L | M | N | O |
|---|---|---|---|---|
| 702 − 18 | 5106 − 1276 | 905 − 125 | 380 + 280 | 5276 − 4506 |

# Lesson 35

| Facts | + | Problems | + | Bonus | = | TOTAL |
|---|---|---|---|---|---|---|
| | | | | | | |

## 1

| 9 | 9 | 9 | 9 | 9 | 9 | 9 | 9 |
|---|---|---|---|---|---|---|---|
| − 3 | − 1 | − 4 | − 2 | − 0 | − 4 | − 2 | − 3 |

## 2

| 13 | 16 | 5 | 13 | 5 | 13 | 16 | 13 |
|---|---|---|---|---|---|---|---|
| − 5 | − 7 | − 3 | − 8 | − 2 | − 5 | − 7 | − 8 |

## 3

**A** 3 hammers, how many tools, 4 saws

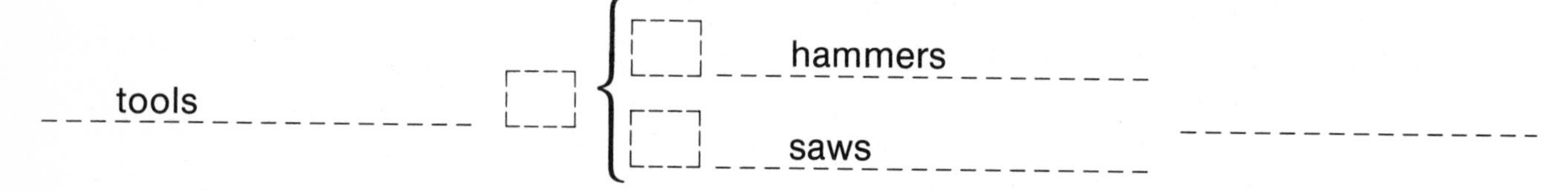


**B** 3 girls, 5 children, how many boys

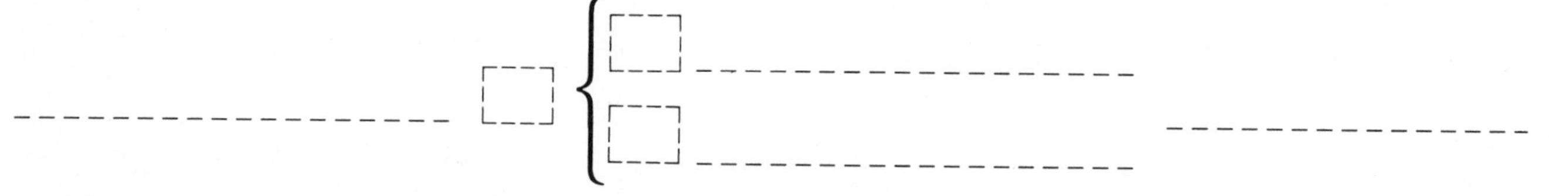

**C** 7 pens, how many red pens, 3 blue pens

**D** 5 pets, 4 dogs, how many cats

**E** 5 big boys, how many boys, 3 little boys

## 4

| | | | | | | | |
|---|---|---|---|---|---|---|---|
| 10 − 4 | 10 − 9 | 10 − 6 | 10 − 3 | 10 − 5 | 10 − 7 | 10 − 1 | 10 − 8 |
| 10 − 6 | 10 − 2 | 10 − 5 | 10 − 7 | 10 − 9 | 10 − 4 | 10 − 8 | 10 − 1 |

## 5

| | | | | | | | |
|---|---|---|---|---|---|---|---|
| 16 − 7 | 14 − 9 | 12 − 5 | 14 − 8 | 12 − 6 | 14 − 6 | 12 − 7 | 8 − 3 |
| 14 − 9 | 7 − 3 | 8 − 5 | 7 − 5 | 13 − 9 | 16 − 7 | 12 − 4 | 14 − 6 |
| 14 − 8 | 16 − 9 | 7 − 4 | 16 − 7 | 11 − 9 | 14 − 7 | 17 − 9 | 8 − 3 |
| 10 − 2 | 8 − 5 | 10 − 4 | 13 − 9 | 16 − 8 | 12 − 9 | 11 − 9 | 15 − 9 |

## 6

**A**
Dr. Holzburg examined 7 people in their hospital rooms. Then she treated 3 more people in the emergency room. How many people did Dr. Holzburg take care of at the hospital?

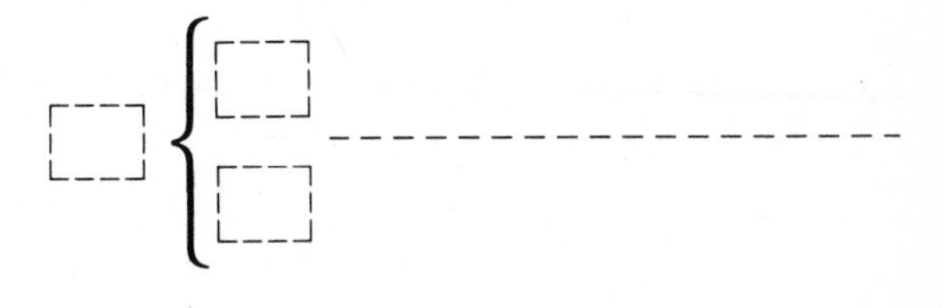

**B**
Pierre and his father built 8 birdhouses. They sold 3 of them. How many birdhouses do they have left?

Part 6 continues on the next page.

**C**
Mrs. Martin is a bus driver. There are 5 people on her bus. She picks up 5 people. How many people are on the bus now?

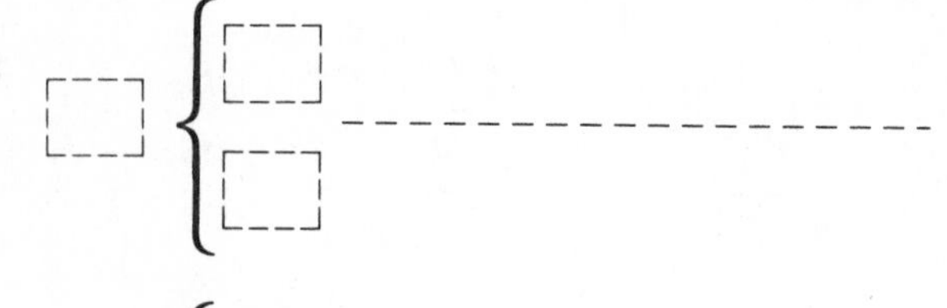

**D**
Audrey took 6 photographs. Then she took 4 more. How many photographs has she taken now?

**E**
Samantha wrote a play. 7 friends read her play. 5 more friends read her play. How many friends in all read Samantha's play?

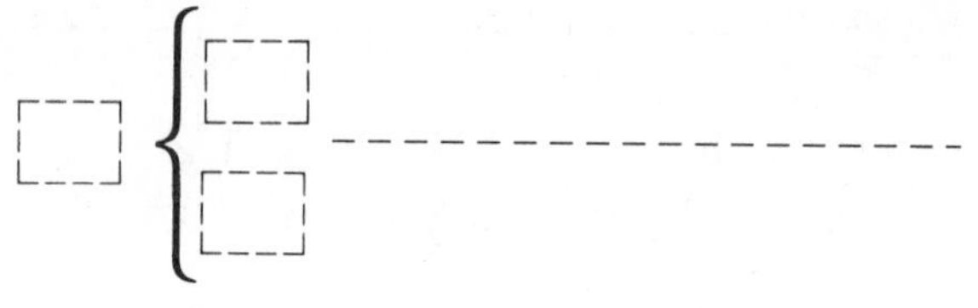

**F**
Barbara's basketball team played 13 games at their school. They played 6 games at other schools. How many games did they play in all?

## 7

**A** $\begin{array}{r} 956 \\ -368 \\ \hline \end{array}$

**B** $\begin{array}{r} 6280 \\ +2084 \\ \hline \end{array}$

**C** $\begin{array}{r} 8042 \\ +\ 160 \\ \hline \end{array}$

**D** $\begin{array}{r} 9520 \\ -9480 \\ \hline \end{array}$

**E** $\begin{array}{r} 802 \\ -315 \\ \hline \end{array}$

**F** $\begin{array}{r} 5328 \\ -\ 718 \\ \hline \end{array}$

**G** $\begin{array}{r} 3412 \\ -\ 807 \\ \hline \end{array}$

**H** $\begin{array}{r} 6252 \\ -6188 \\ \hline \end{array}$

**I** $\begin{array}{r} 6204 \\ +1804 \\ \hline \end{array}$

**J** $\begin{array}{r} 904 \\ -\ 20 \\ \hline \end{array}$

**K** $\begin{array}{r} 72 \\ -14 \\ \hline \end{array}$

**L** $\begin{array}{r} 7062 \\ -3258 \\ \hline \end{array}$

**M** $\begin{array}{r} 7236 \\ +\ 836 \\ \hline \end{array}$

**N** $\begin{array}{r} 9307 \\ -3692 \\ \hline \end{array}$

**O** $\begin{array}{r} 3642 \\ -\ 802 \\ \hline \end{array}$

# Lesson 36

| Facts | + | Problems | + | Bonus | = | TOTAL |
|---|---|---|---|---|---|---|
| | | | | | | |

## 1

| 9 | 9 | 9 | 9 | 9 | 9 | 9 | 9 |
|---|---|---|---|---|---|---|---|
| − 4 | − 2 | − 3 | − 4 | − 1 | − 3 | − 2 | − 2 |

## 2

| 13 | 13 | 13 | 13 | 13 | 13 | 13 | 13 |
|---|---|---|---|---|---|---|---|
| − 9 | − 5 | − 7 | − 8 | − 5 | − 7 | − 9 | − 8 |

## 3

| A | B | C | D | E |
|---|---|---|---|---|
| 279 | 146 | 657 | 328 | 864 |

## 4

**A** 8 houses, how many buildings, 2 stores

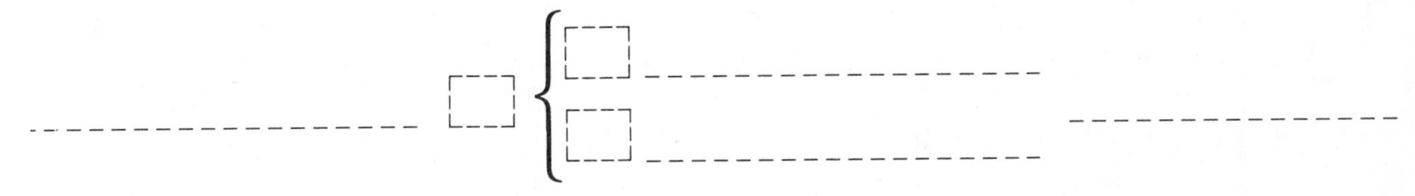

**B** 5 dogs, 9 pets, how many cats

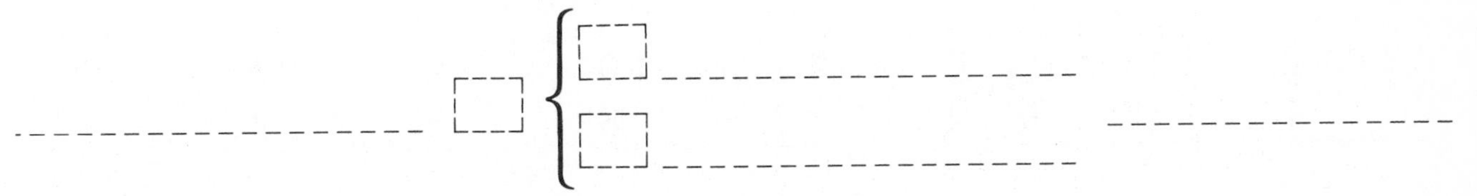

**C** how many chairs, 2 wooden chairs, 7 plastic chairs

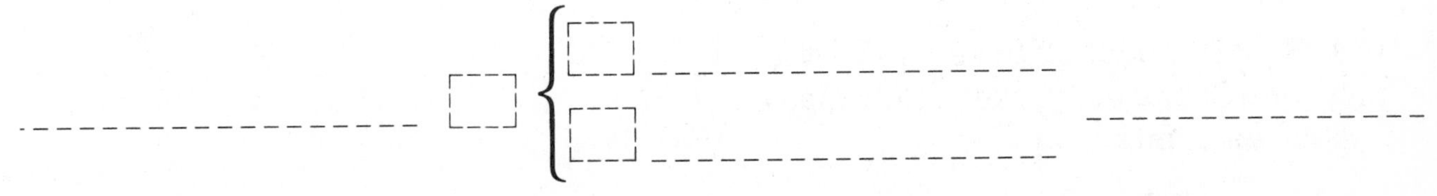

**D** 10 tools, 3 saws, how many hammers

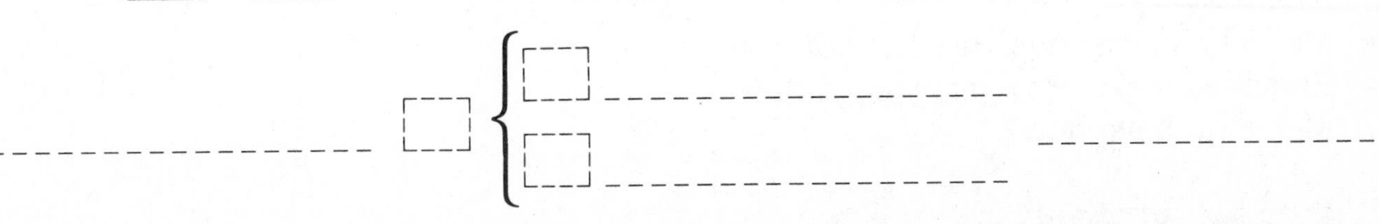

Part 4 continues on the next page.

**E** 4 girls, how many boys, 7 children

**F** 5 happy children, 3 sad children, how many children

## 5

| | | | | | | | |
|---|---|---|---|---|---|---|---|
| 10 − 7 | 16 − 7 | 5 − 2 | 10 − 6 | 14 − 9 | 5 − 3 | 10 − 8 | 16 − 9 |
| 10 − 5 | 11 − 9 | 10 − 4 | 14 − 9 | 10 − 3 | 12 − 8 | 12 − 9 | 12 − 3 |
| 12 − 6 | 10 − 2 | 18 − 9 | 10 − 6 | 13 − 9 | 12 − 7 | 10 − 7 | 7 − 3 |
| 12 − 4 | 15 − 9 | 10 − 1 | 12 − 3 | 10 − 8 | 12 − 5 | 17 − 9 | 8 − 3 |

## 6

**A** Mrs. Benally bought 8 tickets to visit a museum. She used 2 tickets. How many tickets were left?

☐ tickets

**B** Mr. Harjo had 5 eggs. He used 3 of them to make a cake. How many eggs does Mr. Harjo have now?

☐

Part 6 continues on the next page.

**C** Rita had 2 gold coins. Then she received 3 more gold coins for her birthday. How many gold coins did Rita end up with?

[ ] ____________________

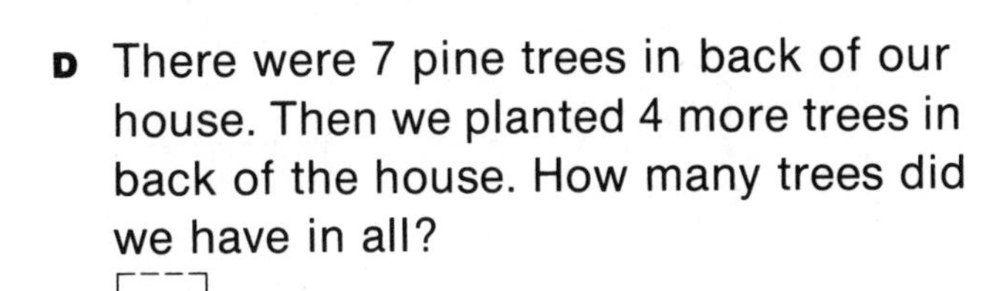

**D** There were 7 pine trees in back of our house. Then we planted 4 more trees in back of the house. How many trees did we have in all?

[ ] ____________________

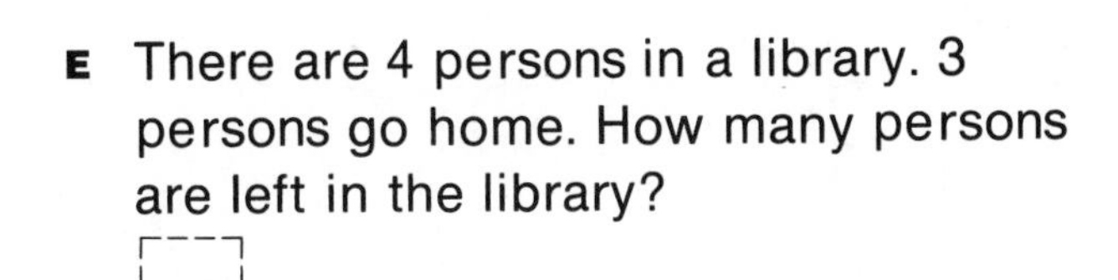

**E** There are 4 persons in a library. 3 persons go home. How many persons are left in the library?

[ ] ____________________

## 7

**A** $\begin{array}{r} 7307 \\ -\ \ 913 \\ \hline \end{array}$

**B** $\begin{array}{r} 3260 \\ -2752 \\ \hline \end{array}$

**C** $\begin{array}{r} 3846 \\ +\ \ 256 \\ \hline \end{array}$

**D** $\begin{array}{r} 4826 \\ +\ \ \ \ 88 \\ \hline \end{array}$

**E** $\begin{array}{r} 8084 \\ -4557 \\ \hline \end{array}$

**F** $\begin{array}{r} 7108 \\ -1398 \\ \hline \end{array}$

**G** $\begin{array}{r} 804 \\ -316 \\ \hline \end{array}$

**H** $\begin{array}{r} 7582 \\ -7483 \\ \hline \end{array}$

**I** $\begin{array}{r} 604 \\ -\ \ 20 \\ \hline \end{array}$

**J** $\begin{array}{r} 706 \\ -589 \\ \hline \end{array}$

**K** $\begin{array}{r} 280 \\ +280 \\ \hline \end{array}$

**L** $\begin{array}{r} 50 \\ -23 \\ \hline \end{array}$

**M** $\begin{array}{r} 5162 \\ -4180 \\ \hline \end{array}$

**N** $\begin{array}{r} 3046 \\ -2320 \\ \hline \end{array}$

**O** $\begin{array}{r} 60 \\ -35 \\ \hline \end{array}$

# Lesson 37

Test + Facts + Problems + Bonus = TOTAL

## 1

| 13 | 13 | 13 | 13 | 13 | 13 | 13 | 13 |
|---|---|---|---|---|---|---|---|
| − 7 | − 5 | − 8 | − 7 | − 8 | − 5 | − 7 | − 9 |

## 2

**A** 9 { 4 ____ / [ ] ____

**B** 9 { 2 ____ / [ ] ____

**C** 9 { 3 ____ / [ ] ____

**D** 9 { 1 ____ / [ ] ____

## 3

| 9 | 9 | 9 | 9 | 9 | 9 | 9 | 9 |
|---|---|---|---|---|---|---|---|
| − 4 | − 2 | − 7 | − 3 | − 6 | − 4 | − 5 | − 3 |

| 9 | 9 | 9 | 9 | 9 | 9 | 9 | 9 |
|---|---|---|---|---|---|---|---|
| − 6 | − 4 | − 8 | − 5 | − 7 | − 3 | − 4 | − 6 |

## 4

| 5 | 16 | 10 | 16 | 5 | 11 | 10 | 14 |
|---|---|---|---|---|---|---|---|
| − 2 | − 7 | − 6 | − 9 | − 3 | − 9 | − 7 | − 9 |

| 13 | 10 | 7 | 12 | 8 | 17 | 10 | 7 |
|---|---|---|---|---|---|---|---|
| − 9 | − 3 | − 2 | − 4 | − 3 | − 9 | − 8 | − 5 |

| 7 | 8 | 12 | 15 | 12 | 7 | 18 | 10 |
|---|---|---|---|---|---|---|---|
| − 4 | − 5 | − 3 | − 9 | − 5 | − 3 | − 9 | − 6 |

Part 4 continues on the next page.

| 12 | 10 | 12 | 10 | 12 | 10 | 12 | 12 |
|---|---|---|---|---|---|---|---|
| − 7 | − 4 | − 6 | − 2 | − 8 | − 7 | − 3 | − 9 |

## 5

| A | B | C | D | E |
|---|---|---|---|---|
| 524 | 738 | 916 | 437 | 514 |

## 6

**A** Yasmin and her brother own 3 dogs and 4 cats. They have trained them to do tricks. How many pets have they trained?

**B** Rick was building a tree house for his brother and sister. Rick has 9 tools. He has 4 hammers. The rest are saws. How many saws does he have?

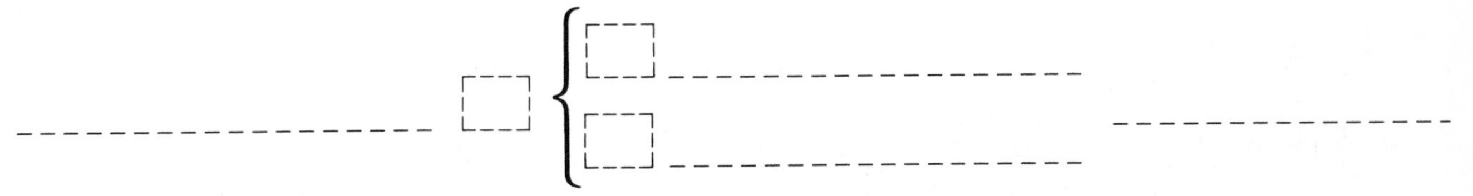

**C** Phyllis has 4 white rabbits. She has 3 brown rabbits. How many rabbits does Phyllis have?

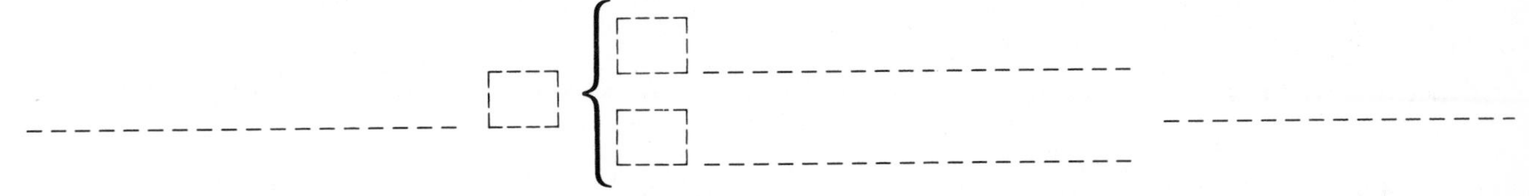

**D** There are 7 children in a swimming pool. There are 4 girls in the pool. How many boys are in the pool?

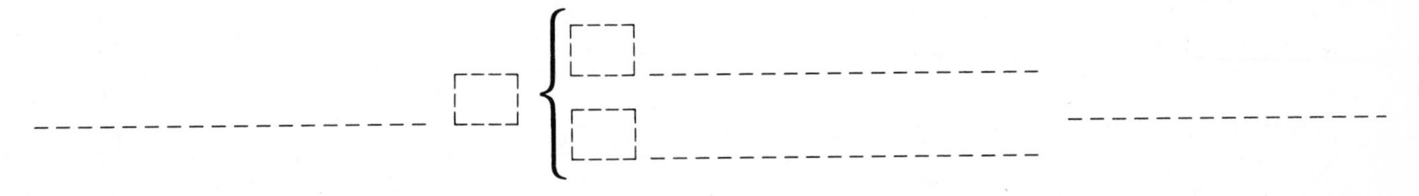

Part 6 continues on the next page.

**E** There are 6 houses on Adams Street. There are 4 shops on Adams Street. How many buildings are on Adams Street?

**F** We put 9 chairs on the stage for a play. There were 4 rocking chairs. The rest were plain wooden chairs. How many plain wooden chairs were there?

## 7

**A** Eric knew how to play 7 songs on his guitar. This month he learned 2 more songs. How many songs can Eric play now?

songs

**B** Tim's jacket had 8 buttons. Tim lost 3 of the buttons. How many buttons were left on Tim's jacket?

**C** There were 9 children playing in the park. 4 children went home. How many children were still in the park?

**D** Bill has built 8 model planes. If he builds 3 more model planes, how many planes will he have?

**8**

**A**
$$\begin{array}{r} 5303 \\ -\ 1949 \\ \hline \end{array}$$

**B**
$$\begin{array}{r} 6502 \\ -\ \ 786 \\ \hline \end{array}$$

**C**
$$\begin{array}{r} 4704 \\ -\ \ 817 \\ \hline \end{array}$$

**D**
$$\begin{array}{r} 5306 \\ -\ 4688 \\ \hline \end{array}$$

**E**
$$\begin{array}{r} 824 \\ +\ 136 \\ \hline \end{array}$$

**F**
$$\begin{array}{r} 5675 \\ -\ 1832 \\ \hline \end{array}$$

**G**
$$\begin{array}{r} 7306 \\ -\ 6218 \\ \hline \end{array}$$

**H**
$$\begin{array}{r} 5124 \\ -\ 2354 \\ \hline \end{array}$$

**I**
$$\begin{array}{r} 1936 \\ +\ \ 236 \\ \hline \end{array}$$

**J**
$$\begin{array}{r} 804 \\ -\ 420 \\ \hline \end{array}$$

**K**
$$\begin{array}{r} 3248 \\ -\ \ 701 \\ \hline \end{array}$$

**L**
$$\begin{array}{r} 96 \\ -\ 30 \\ \hline \end{array}$$

**M**
$$\begin{array}{r} 5362 \\ +\ 1682 \\ \hline \end{array}$$

**N**
$$\begin{array}{r} 80 \\ -\ 53 \\ \hline \end{array}$$

**O**
$$\begin{array}{r} 6040 \\ -\ 2180 \\ \hline \end{array}$$

**P**
$$\begin{array}{r} 5308 \\ -\ 1248 \\ \hline \end{array}$$

# Lesson 38

| Facts | + | Problems | + | Bonus | = | TOTAL |
|---|---|---|---|---|---|---|
| | | | | | | |

## 1

| | | | | | | | |
|---|---|---|---|---|---|---|---|
| 9 | 9 | 9 | 9 | 9 | 9 | 9 | 9 |
| − 6 | − 7 | − 5 | − 4 | − 7 | − 5 | − 3 | − 6 |

| | | | | | | | |
|---|---|---|---|---|---|---|---|
| 9 | 9 | 9 | 9 | 9 | 9 | 9 | 9 |
| − 4 | − 8 | − 6 | − 4 | − 5 | − 3 | − 2 | − 7 |

## 2

**A**

13 { 7 ________ / [ ] ________

**B**

13 { 9 ________ / [ ] ________

**C**

13 { 8 ________ / [ ] ________

**D**

13 { 10 ________ / [ ] ________

## 3

| | | | | | | | |
|---|---|---|---|---|---|---|---|
| 13 | 13 | 13 | 13 | 13 | 13 | 13 | 13 |
| − 4 | − 8 | − 9 | − 6 | − 7 | − 9 | − 5 | − 4 |

| | | | | | | | |
|---|---|---|---|---|---|---|---|
| 13 | 13 | 13 | 13 | 13 | 13 | 13 | 13 |
| − 6 | − 9 | − 8 | − 5 | − 4 | − 7 | − 9 | − 4 |

## 4

| | | | | | | | |
|---|---|---|---|---|---|---|---|
| 16 | 10 | 7 | 16 | 10 | 11 | 7 | 10 |
| − 7 | − 6 | − 2 | − 9 | − 7 | − 9 | − 5 | − 8 |

| | | | | | | | |
|---|---|---|---|---|---|---|---|
| 15 | 10 | 12 | 10 | 7 | 10 | 16 | 10 |
| − 9 | − 3 | − 9 | − 7 | − 3 | − 2 | − 9 | − 9 |

Part 4 continues on the next page.

| 14 | 10 | 7 | 10 | 12 | 18 | 8 | 10 |
|---|---|---|---|---|---|---|---|
| − 7 | − 8 | − 4 | − 4 | − 6 | − 9 | − 4 | − 6 |

| 17 | 10 | 6 | 14 | 10 | 8 | 13 | 12 |
|---|---|---|---|---|---|---|---|
| − 9 | − 5 | − 3 | − 9 | − 9 | − 8 | − 9 | − 6 |

## 5

**A**
824 − 13 = ______

824
− 13

**B**
627 − 9 = ______

**C**
302 − 37 = ______

## 6

**A** Miss Soto is a barber. She gave 7 adult haircuts. She also gave 4 children's haircuts. How many haircuts did Miss Soto give?

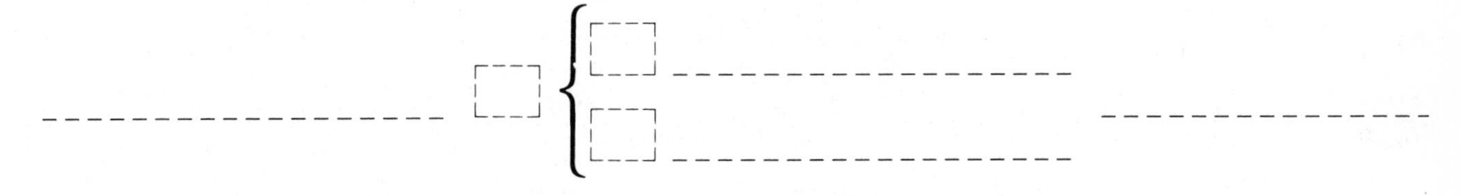

**B** In our school there are 10 classrooms. There are 4 large classrooms. How many small classrooms are there?

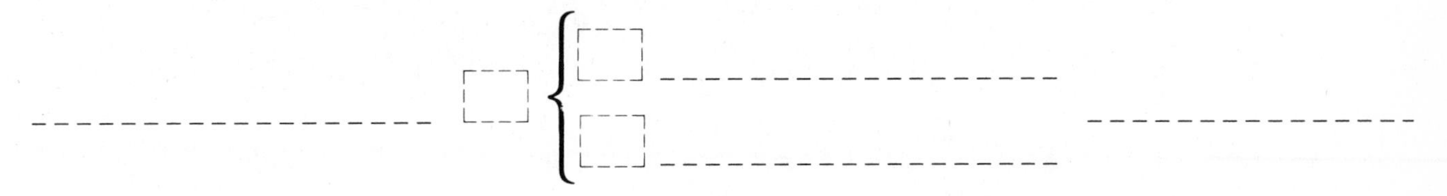

**C** 10 children are playing tennis in the park. There are 7 boys. How many girls are there?

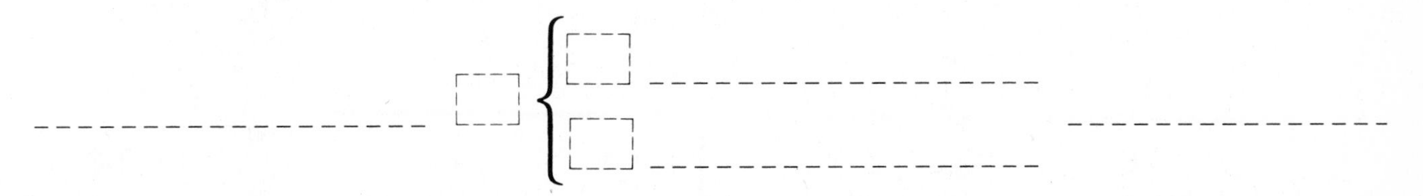

Part 6 continues on the next page.

**D** Jaime was giving a party. On Sunday he invited 10 boys. On Monday he invited 6 girls. How many children did Jaime invite to his party?

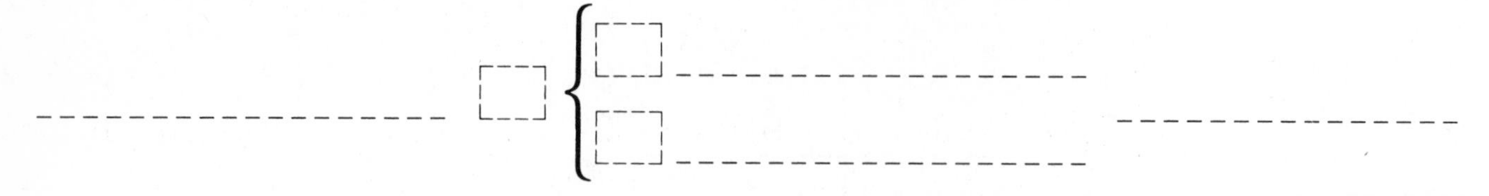

**E** Jack washed and dried 5 new plates. Then he washed and dried 3 old plates. How many plates did Jack wash and dry?

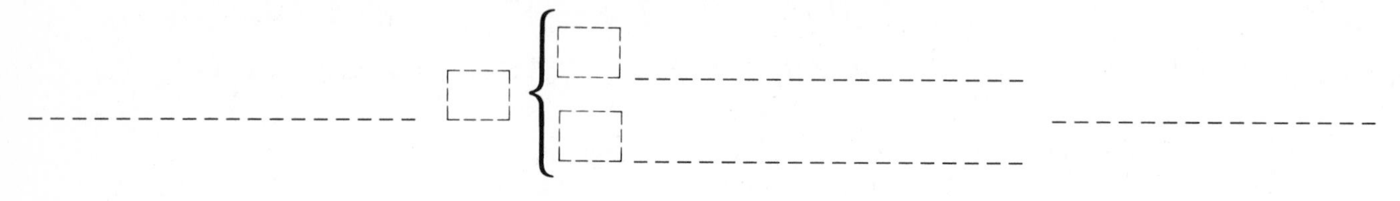

**F** There were 9 birds on a park bench. 4 pigeons flew off the bench. Only crows are left. How many crows are left on the bench?

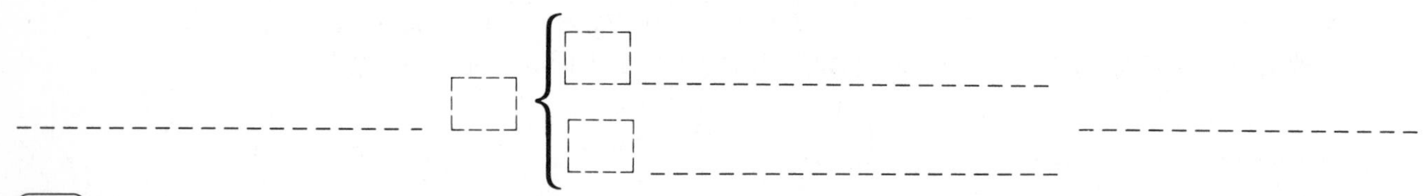

## 7

**A** $4378 - 3520$

**B** $8306 - 3541$

**C** $8052 - 78$

**D** $3280 - 180$

**E** $802 - 98$

**F** $3502 - 1786$

**G** $6280 + 1004$

**H** $938 + 27$

**I** $4125 - 1285$

**J** $9028 - 8357$

**K** $7248 - 1761$

**L** $3215 - 710$

**M** $5482 - 4859$

**N** $403 - 33$

**O** $8506 - 798$

8

A Paul made 7 rugs. He sold 4 rugs at an art fair. How many rugs did he have left?

☐ _________________

B Mr. Duval bought a parrot that knew 9 words. Mr. Duval taught the parrot 5 words. How many words could the parrot say then?

☐ _________________

C Mr. Wells raises a bridge so tall boats can pass through. He raised the bridge 9 times on Saturday. He also raised the bridge 9 times on Sunday. How many times did Mr. Wells raise the bridge?

☐ _________________

D After our school picnic, Emily and some friends filled 9 bags with litter from the play area. Then they filled 2 bags with litter from tables. How many bags of litter did Emily and her friends collect?

☐ _________________

# Lesson 39

| Facts | + | Problems | + | Bonus | = | TOTAL |
|---|---|---|---|---|---|---|
| | | | | | | |

**1**

| | | | | | | | |
|---|---|---|---|---|---|---|---|
| 9 − 6 | 9 − 7 | 9 − 4 | 9 − 5 | 9 − 8 | 9 − 3 | 9 − 6 | 9 − 7 |
| 9 − 2 | 9 − 0 | 9 − 5 | 9 − 1 | 9 − 6 | 9 − 2 | 9 − 7 | 9 − 4 |

**2**

**A** 13 { 7 ______ / ☐ ______

**B** 13 { 9 ______ / ☐ ______

**C** 13 { 8 ______ / ☐ ______

**D** 13 { 10 ______ / ☐ ______

**3**

| | | | | | | | |
|---|---|---|---|---|---|---|---|
| 13 − 4 | 13 − 9 | 13 − 5 | 13 − 8 | 13 − 4 | 13 − 6 | 13 − 7 | 13 − 9 |
| 13 − 7 | 13 − 4 | 13 − 8 | 13 − 4 | 13 − 5 | 13 − 6 | 13 − 7 | 13 − 8 |

**4**

| | | | | | | | |
|---|---|---|---|---|---|---|---|
| 16 − 7 | 16 − 9 | 13 − 9 | 5 − 2 | 11 − 9 | 14 − 6 | 8 − 3 | 14 − 9 |
| 10 − 6 | 15 − 9 | 10 − 8 | 12 − 3 | 10 − 6 | 17 − 9 | 10 − 7 | 13 − 9 |

Part 4 continues on the next page.

$$\begin{array}{r} 12 \\ -\ 9 \\ \hline \end{array} \quad \begin{array}{r} 12 \\ -\ 5 \\ \hline \end{array} \quad \begin{array}{r} 10 \\ -\ 8 \\ \hline \end{array} \quad \begin{array}{r} 18 \\ -\ 9 \\ \hline \end{array} \quad \begin{array}{r} 10 \\ -\ 4 \\ \hline \end{array} \quad \begin{array}{r} 10 \\ -\ 8 \\ \hline \end{array} \quad \begin{array}{r} 12 \\ -\ 3 \\ \hline \end{array} \quad \begin{array}{r} 12 \\ -\ 8 \\ \hline \end{array}$$

## 5

**A**
$4208 - 60 =$ ______

**B**
$904 - 7 =$ ______

**C**
$5154 - 148 =$ ______

## 6

**A** Ivan went to the market and bought 6 carrots. He also bought 4 onions. How many vegetables did Ivan buy?

______ ☐ { ☐ ______ ; ☐ ______ } ______

**B** A science class was studying birds. On a trip to the mountains the students saw 8 birds. 7 were hawks. The rest were eagles. How many eagles did they see?

______ ☐ { ☐ ______ ; ☐ ______ } ______

**C** Our team had 7 blue uniforms. The rest of the uniforms were red. There were 9 uniforms in all. How many red uniforms were there?

______ ☐ { ☐ ______ ; ☐ ______ } ______

Part 6 continues on the next page.

**D** Kitty and her friends wash cars on weekends. One Saturday Kitty washed 5 cars. 3 of them were new. How many old cars did she wash?

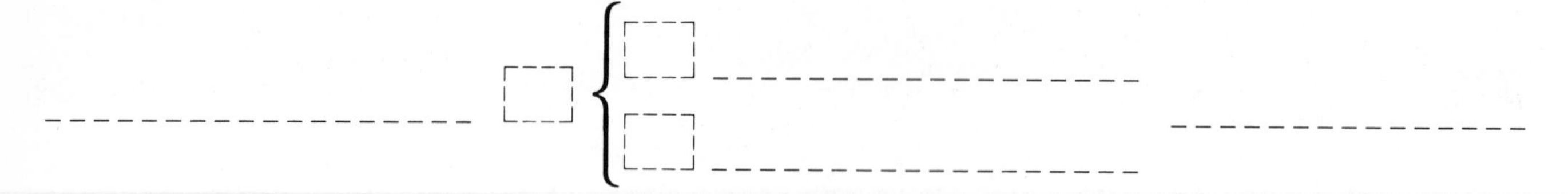

**E** Every Sunday 6 people come to the tennis courts. 4 women play tennis on one court. Men play on the other court. How many men play on the other court?

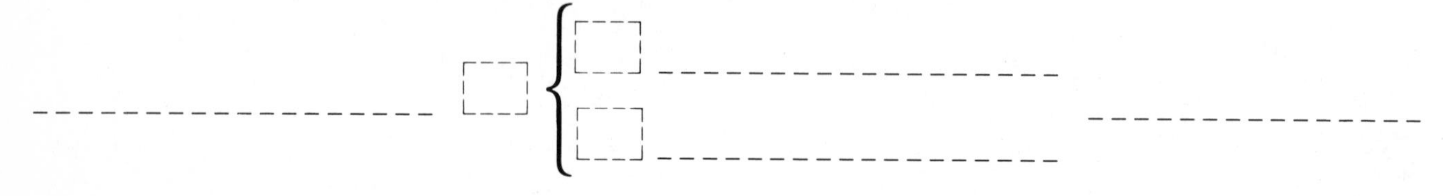

**F** Fumiko put out birdseed. He watched as 9 blackbirds flew down to eat. Next 9 robins arrived. How many birds ate seed?

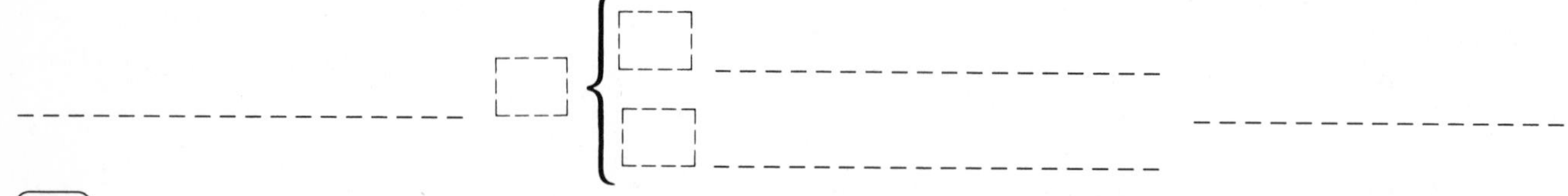

## 7

**A** $828 + 80$

**B** $6308 - 3889$

**C** $9246 - 86$

**D** $406 - 355$

**E** $802 - 376$

**F** $4082 - 3092$

**G** $4268 - 608$

**H** $3642 + 1082$

**I** $3304 - 617$

**J** $6047 - 3442$

**K** $3826 + 800$

**L** $3045 + 2955$

**M** $6468 - 2580$

**N** $7045 - 6182$

**O** $4304 - 687$

## 8

**A** There were 10 children at a party. 6 of the children went home early. How many children were still at the party?

[ ] ____________________

**B** Sophia had 10 model cars. She built 6 more model cars. How many cars does she have now?

[ ] ____________________

**C** If you have 7 books and your friend gives you 4 books, how many books will you have?

[ ] ____________________

**D** Stella had 9 apples. She gave 3 apples to her horse. How many apples does Stella have left?

[ ] ____________________

# Lesson 40

| Facts | + | Problems | + | Bonus | = | TOTAL |
|---|---|---|---|---|---|---|
| | | | | | | |

## 1

| 13 | 13 | 13 | 13 | 13 | 13 | 13 | 13 |
|---|---|---|---|---|---|---|---|
| − 4 | − 7 | − 6 | − 5 | − 9 | − 7 | − 4 | − 6 |

| 13 | 13 | 13 | 13 | 13 | 13 | 13 | 13 |
|---|---|---|---|---|---|---|---|
| − 5 | − 8 | − 4 | − 8 | − 6 | − 9 | − 7 | − 4 |

## 2

**A**

9 { 3 ______ ; [ ] ______

**B**

9 { 2 ______ ; [ ] ______

**C**

9 { 4 ______ ; [ ] ______

**D**

9 { 1 ______ ; [ ] ______

## 3

| 9 | 9 | 9 | 9 | 9 | 9 | 9 | 9 |
|---|---|---|---|---|---|---|---|
| − 5 | − 7 | − 6 | − 3 | − 5 | − 4 | − 7 | − 2 |

| 9 | 9 | 9 | 9 | 9 | 9 | 9 | 9 |
|---|---|---|---|---|---|---|---|
| − 4 | − 6 | − 8 | − 2 | − 7 | − 1 | − 5 | − 3 |

## 4

| 10 | 16 | 12 | 10 | 16 | 11 | 10 | 15 |
|---|---|---|---|---|---|---|---|
| − 5 | − 7 | − 7 | − 4 | − 9 | − 9 | − 7 | − 9 |

| 12 | 13 | 14 | 12 | 10 | 17 | 8 | 10 |
|---|---|---|---|---|---|---|---|
| − 4 | − 9 | − 7 | − 3 | − 6 | − 9 | − 3 | − 3 |

Part 4 continues on the next page.

| | | | | | | | |
|---|---|---|---|---|---|---|---|
| $10 - 2$ | $10 - 5$ | $18 - 9$ | $10 - 2$ | $16 - 7$ | $14 - 9$ | $10 - 8$ | $10 - 4$ |
| $7 - 5$ | $10 - 3$ | $12 - 3$ | $12 - 5$ | $10 - 9$ | $8 - 5$ | $12 - 8$ | $7 - 3$ |

**A** The volleyball team had 7 children. 4 were girls. How many were boys?

☐ ________________

**B** Jorge wrote 8 songs in all last winter. He wrote 7 sad songs. The rest were funny songs. How many funny songs did he write?

☐ ________________

**C** At a country fair, Joan sold 10 peaches. Then she sold 5 pears. How many pieces of fruit did Joan sell?

☐ ________________

**D** Mrs. Stein owns a yarn shop. Last winter she bought 7 big boxes of yarn. She also bought 4 small boxes of yarn. How many boxes of yarn did she buy?

☐ ________________

**E** In the park, 4 men were jogging. There were 9 people jogging in the park. How many women were jogging?

☐ ________________

**F** Pam collected shells at the beach. She found 4 gray shells on Monday. On Tuesday she found 10 white shells. How many shells in all did Pam find?

☐ ________________

**A** There were 532 children in our school. 18 children moved to another school. How many children are in our school now?

$$\begin{array}{r} 532 \\ -\ \ 18 \\ \hline \end{array}$$

**B** Mr. Ramirez is a plant scientist. Last year he found 2431 plants. This year he found 92 more plants. How many plants does Mr. Ramirez have now?

**C** Miss Tanaka owns a gas station. She had 8240 liters of gas. She sold 6105 liters of gas. How many liters does she have now?

## 7

**A** $\begin{array}{r} 802 \\ -796 \\ \hline \end{array}$ **B** $\begin{array}{r} 69 \\ +60 \\ \hline \end{array}$ **C** $\begin{array}{r} 6066 \\ -3919 \\ \hline \end{array}$ **D** $\begin{array}{r} 3502 \\ -\ \ 786 \\ \hline \end{array}$ **E** $\begin{array}{r} 9654 \\ +\ \ 423 \\ \hline \end{array}$

**F** $\begin{array}{r} 430 \\ -284 \\ \hline \end{array}$ **G** $\begin{array}{r} 8345 \\ +1245 \\ \hline \end{array}$ **H** $\begin{array}{r} 7042 \\ -3638 \\ \hline \end{array}$ **I** $\begin{array}{r} 4036 \\ -\ \ \ \ 81 \\ \hline \end{array}$ **J** $\begin{array}{r} 4130 \\ -\ \ \ \ \ \ 9 \\ \hline \end{array}$

**K** $\begin{array}{r} 5103 \\ -4941 \\ \hline \end{array}$ **L** $\begin{array}{r} 4506 \\ -\ \ 798 \\ \hline \end{array}$ **M** $\begin{array}{r} 4310 \\ -\ \ 900 \\ \hline \end{array}$ **N** $\begin{array}{r} 3042 \\ -1998 \\ \hline \end{array}$ **O** $\begin{array}{r} 486 \\ +\ \ 16 \\ \hline \end{array}$

# Lesson 41

Facts + Problems + Bonus = TOTAL

## 1

| 9 | 9 | 9 | 9 | 9 | 9 | 9 | 9 |
|---|---|---|---|---|---|---|---|
| − 7 | − 4 | − 3 | − 6 | − 2 | − 8 | − 7 | − 1 |

| 9 | 9 | 9 | 9 | 9 | 9 | 9 | 9 |
|---|---|---|---|---|---|---|---|
| − 5 | − 6 | − 4 | − 7 | − 3 | − 6 | − 2 | − 7 |

## 2

**A**

13 { 5 ______ ; ☐ ______

**B**

13 { 6 ______ ; ☐ ______

**C**

13 { 9 ______ ; ☐ ______

**D**

13 { 10 ______ ; ☐ ______

## 3

| 13 | 13 | 13 | 13 | 13 | 13 | 13 | 13 |
|---|---|---|---|---|---|---|---|
| − 7 | − 9 | − 4 | − 8 | − 5 | − 6 | − 4 | − 7 |

| 13 | 13 | 13 | 13 | 13 | 13 | 13 | 13 |
|---|---|---|---|---|---|---|---|
| − 6 | − 8 | − 4 | − 5 | − 9 | − 7 | − 6 | − 8 |

## 4

| 10 | 16 | 10 | 14 | 10 | 16 | 10 | 11 |
|---|---|---|---|---|---|---|---|
| − 6 | − 9 | − 7 | − 9 | − 4 | − 7 | − 3 | − 9 |

| 8 | 10 | 12 | 7 | 12 | 16 | 10 | 17 |
|---|---|---|---|---|---|---|---|
| − 3 | − 6 | − 3 | − 3 | − 4 | − 9 | − 2 | − 9 |

Part 4 continues on the next page.

$10 - 8$ | $14 - 9$ | $10 - 3$ | $8 - 5$ | $15 - 9$ | $7 - 5$ | $12 - 3$ | $12 - 7$

$12 - 3$ | $14 - 6$ | $16 - 8$ | $10 - 3$ | $7 - 2$ | $10 - 7$ | $14 - 8$ | $7 - 5$

## 5

**A** $5102 - 87$

**B** $6104 - 268$

**C** $3106 - 418$

**D** $7104 - 237$

## 6

**A** Our school had a dog show for German shepherds and hunting dogs. There were 15 German shepherds in the show. 19 dogs had been brought to the show. How many hunting dogs were there?

☐ ________________

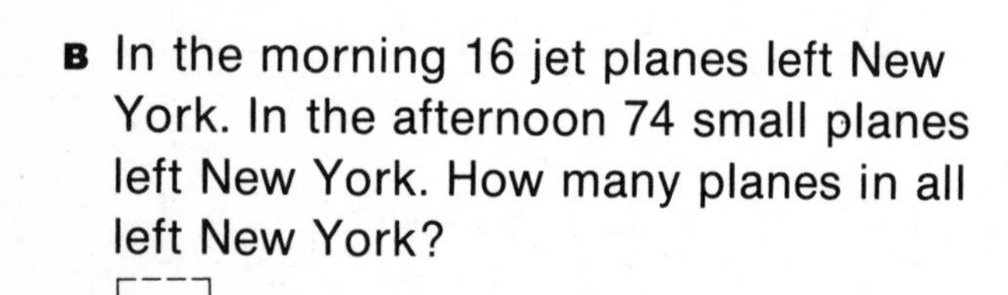

**B** In the morning 16 jet planes left New York. In the afternoon 74 small planes left New York. How many planes in all left New York?

☐ ________________

**C** There are 7 rosebushes in our garden. We have 4 red rosebushes. The rest are yellow. How many yellow rosebushes do we have?

☐ ________________

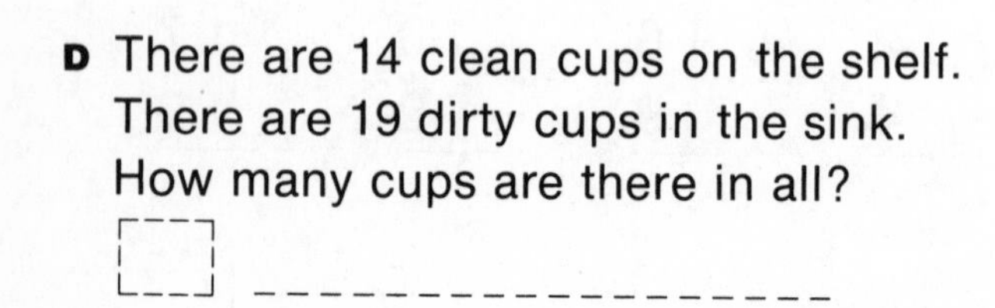

**D** There are 14 clean cups on the shelf. There are 19 dirty cups in the sink. How many cups are there in all?

☐ ________________

Part 6 continues on the next page.

**E** 14 women were in the play. 29 people were in the play. How many men were in the play?

**F** My aunt has 28 white flowers. She has 39 flowers in all. The rest of the flowers are pink. How many pink flowers does my aunt have?

## 7

**A** A pet shop had 186 birds. The shop sold 48 birds. How many birds does the shop have now?

**B** Last month Mr. Vanderpool sold 3184 loaves of bread at his bakery. This month Mr. Vanderpool sold 42 more loaves of bread. How many loaves has he sold?

## 8

| A | B | C | D | E |
|---|---|---|---|---|
| $\begin{array}{r} 920 \\ -\ 397 \\ \hline \end{array}$ | $\begin{array}{r} 8506 \\ -\ 6788 \\ \hline \end{array}$ | $\begin{array}{r} 3514 \\ -\ 3439 \\ \hline \end{array}$ | $\begin{array}{r} 8426 \\ +\ 16 \\ \hline \end{array}$ | $\begin{array}{r} 608 \\ -\ 29 \\ \hline \end{array}$ |

| F | G | H | I | J |
|---|---|---|---|---|
| $\begin{array}{r} 7070 \\ -\ 4028 \\ \hline \end{array}$ | $\begin{array}{r} 6059 \\ -\ 1532 \\ \hline \end{array}$ | $\begin{array}{r} 4109 \\ -\ 29 \\ \hline \end{array}$ | $\begin{array}{r} 3704 \\ -\ 886 \\ \hline \end{array}$ | $\begin{array}{r} 3148 \\ +\ 120 \\ \hline \end{array}$ |

# Lesson 42

Facts + Problems + Bonus = TOTAL

## 1

$10 - 2$ $7 - 2$ $11 - 2$ $8 - 2$ $9 - 2$ $7 - 2$ $10 - 2$ $9 - 2$

$8 - 2$ $9 - 2$ $11 - 2$ $10 - 2$ $8 - 2$ $7 - 2$ $11 - 2$ $10 - 2$

## 2

$13 - 4$ $13 - 9$ $13 - 7$ $13 - 4$ $13 - 5$ $13 - 8$ $13 - 6$ $13 - 10$

$13 - 8$ $13 - 4$ $13 - 9$ $13 - 6$ $13 - 5$ $13 - 7$ $13 - 4$ $13 - 6$

## 3

$9 - 4$ $9 - 7$ $14 - 6$ $9 - 6$ $16 - 7$ $10 - 8$ $9 - 3$ $9 - 5$

$10 - 4$ $13 - 9$ $9 - 1$ $14 - 8$ $14 - 9$ $12 - 3$ $9 - 4$ $17 - 9$

$12 - 3$ $9 - 7$ $15 - 9$ $16 - 7$ $11 - 9$ $13 - 9$ $14 - 8$ $9 - 7$

$12 - 5$ $16 - 7$ $12 - 8$ $9 - 8$ $12 - 4$ $14 - 6$ $12 - 3$ $12 - 9$

**A**

$$\begin{array}{r} 3104 \\ -\quad 68 \\ \hline \end{array}$$

**B**

$$\begin{array}{r} 8102 \\ -\quad 46 \\ \hline \end{array}$$

**C**

$$\begin{array}{r} 4102 \\ -\ 538 \\ \hline \end{array}$$

**D**

$$\begin{array}{r} 4102 \\ -\ 586 \\ \hline \end{array}$$

**5**

**A** Bill painted 43 houses. He painted 52 stores. How many buildings did Bill paint?

☐ ________________

**B** In the waiting room of a doctor's office there is a huge fish tank. There are 14 goldfish in the tank and the rest are sunfish. There are 30 fish in all in the tank. How many sunfish are in the tank?

☐ ________________

**C** There are 40 girls in the kindergarten of school. There are 75 children in kindergarten. How many are boys?

☐ ________________

**D** Miss Manos is selling tickets. She sold 31 circus tickets this morning. This afternoon she sold 50 movie tickets. How many tickets in all did she sell?

☐ ________________

**E** Marcia Kabatie has entered many skating contests. 43 were ice-skating contests and the rest were roller-skating ones. She has taken part in 60 contests in all. How many roller-skating contests has she entered?

☐ ________________

## 6

**A**
$$\begin{array}{r} 863 \\ +\ 64 \\ \hline \end{array}$$

**B**
$$\begin{array}{r} 8504 \\ -\ 768 \\ \hline \end{array}$$

**C**
$$\begin{array}{r} 4280 \\ -\ 859 \\ \hline \end{array}$$

**D**
$$\begin{array}{r} 534 \\ -237 \\ \hline \end{array}$$

**E**
$$\begin{array}{r} 9306 \\ -\ 959 \\ \hline \end{array}$$

**F**
$$\begin{array}{r} 9076 \\ -3179 \\ \hline \end{array}$$

**G**
$$\begin{array}{r} 4280 \\ -\ 365 \\ \hline \end{array}$$

**H**
$$\begin{array}{r} 3408 \\ -2614 \\ \hline \end{array}$$

**I**
$$\begin{array}{r} 8241 \\ -\ 848 \\ \hline \end{array}$$

**J**
$$\begin{array}{r} 3504 \\ -2614 \\ \hline \end{array}$$

## 7

**A** Tom feeds 185 chickens on his aunt's farm. His aunt got 15 more chickens. How many chickens will Tom have to feed now?

[ ] ____________________

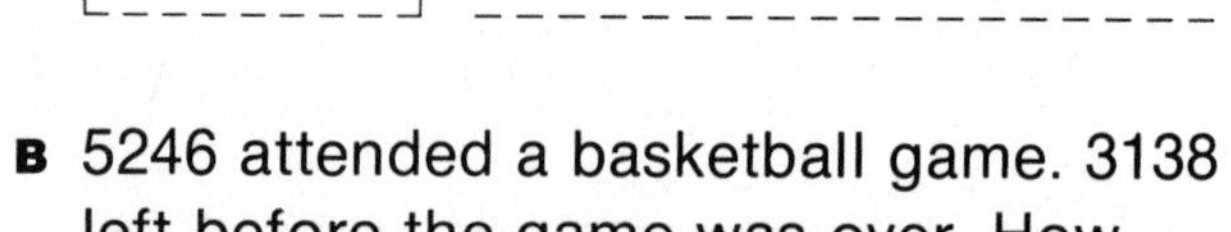

**B** 5246 attended a basketball game. 3138 left before the game was over. How many people stayed for the entire game?

[ ] ____________________

# Lesson 43

| Test | + | Facts | + | Problems | + | Bonus | = | TOTAL |
|---|---|---|---|---|---|---|---|---|
| | | | | | | | | |

## 1

| | | | | | | | |
|---|---|---|---|---|---|---|---|
| 7 | 8 | 6 | 10 | 8 | 5 | 11 | 9 |
| − 2 | − 2 | − 2 | − 2 | − 2 | − 2 | − 2 | − 2 |

| | | | | | | | |
|---|---|---|---|---|---|---|---|
| 5 | 10 | 6 | 11 | 7 | 8 | 4 | 9 |
| − 2 | − 2 | − 2 | − 2 | − 2 | − 2 | − 2 | − 2 |

## 2

| | | | | | | | |
|---|---|---|---|---|---|---|---|
| 13 | 14 | 9 | 13 | 13 | 13 | 9 | 13 |
| − 5 | − 9 | − 3 | − 6 | − 9 | − 8 | − 7 | − 4 |

| | | | | | | | |
|---|---|---|---|---|---|---|---|
| 16 | 13 | 10 | 9 | 13 | 12 | 14 | 9 |
| − 7 | − 7 | − 3 | − 5 | − 5 | − 3 | − 6 | − 5 |

| | | | | | | | |
|---|---|---|---|---|---|---|---|
| 12 | 13 | 13 | 12 | 16 | 9 | 13 | 10 |
| − 8 | − 6 | − 8 | − 3 | − 7 | − 4 | − 4 | − 5 |

| | | | | | | | |
|---|---|---|---|---|---|---|---|
| 14 | 9 | 8 | 13 | 13 | 8 | 10 | 7 |
| − 6 | − 8 | − 3 | − 4 | − 7 | − 5 | − 4 | − 3 |

## 3

| | | | | | | | |
|---|---|---|---|---|---|---|---|
| 15 | 15 | 15 | 15 | 15 | 15 | 15 | 15 |
| − 7 | − 8 | − 10 | − 8 | − 9 | − 7 | − 10 | − 8 |

## 4

**A** Felipe is a bird-watcher for the junior science club. Last fall Felipe saw 15 woodpeckers. He also saw 64 robins. How many birds did Felipe see?

[ ] ____________________

Part 4 continues on the next page.

**B** 114 boys were in a children's hospital. There were 218 children in the hospital. How many girls were in the hospital?

☐ ________________

**C** A jewelry shop sold 14 silver watches in one month. The rest of the watches that were sold were gold. Customers bought 25 watches that month. How many gold watches did the shop sell?

☐ ________________

**D** Mel offered to wash the big windows in his house. His house has 15 small windows. There are 26 windows altogether. How many big windows are there?

☐ ________________

**E** Tamara finished painting a large building. She used 14 yellow cans of paint. She used 20 blue cans of paint. How many cans of paint did Tamara use in all?

☐ ________________

## 5

| A | B | C | D | E |
|---|---|---|---|---|
| $\begin{array}{r} 800 \\ -154 \\ \hline \end{array}$ | $\begin{array}{r} 900 \\ -180 \\ \hline \end{array}$ | $\begin{array}{r} 8500 \\ -7145 \\ \hline \end{array}$ | $\begin{array}{r} 9700 \\ -7360 \\ \hline \end{array}$ | $\begin{array}{r} 800 \\ -142 \\ \hline \end{array}$ |

## 6

| A | B | C | D | E |
|---|---|---|---|---|
| $\begin{array}{r} 5106 \\ -\ \ 468 \\ \hline \end{array}$ | $\begin{array}{r} 970 \\ -734 \\ \hline \end{array}$ | $\begin{array}{r} 864 \\ +132 \\ \hline \end{array}$ | $\begin{array}{r} 4104 \\ -\ \ \ \ 27 \\ \hline \end{array}$ | $\begin{array}{r} 4960 \\ -3899 \\ \hline \end{array}$ |

Part 6 continues on the next page.

| F | G | H | I | J |
|---|---|---|---|---|
| 5140 | 6028 | 8109 | 9304 | 8605 |
| − 98 | + 978 | − 239 | −8949 | − 554 |

## 7

A In a wildlife park, rangers put tags on the legs of 4380 birds. Later, the rangers found that 90 of the birds had flown away. How many were left?

[ ] ________________

B There were 135 children in the swimming pool. 15 children jumped into the pool. How many children were in the pool then?

[ ] ________________

C Sally had 34 school friends. Over the years 19 of her friends have moved away. How many friends are left?

[ ] ________________

D A store had 1485 coats. It sold 665 coats. How many are left?

[ ] ________________

# Lesson 44

| Test | + | Facts | + | Problems | + | Bonus | = | TOTAL |
|---|---|---|---|---|---|---|---|---|
| | | | | | | | | |

## 1

$\begin{array}{r}9\\-2\\\hline\end{array}$ $\begin{array}{r}6\\-2\\\hline\end{array}$ $\begin{array}{r}8\\-2\\\hline\end{array}$ $\begin{array}{r}5\\-2\\\hline\end{array}$ $\begin{array}{r}10\\-2\\\hline\end{array}$ $\begin{array}{r}3\\-2\\\hline\end{array}$ $\begin{array}{r}7\\-2\\\hline\end{array}$ $\begin{array}{r}11\\-2\\\hline\end{array}$

$\begin{array}{r}8\\-2\\\hline\end{array}$ $\begin{array}{r}4\\-2\\\hline\end{array}$ $\begin{array}{r}9\\-2\\\hline\end{array}$ $\begin{array}{r}7\\-2\\\hline\end{array}$ $\begin{array}{r}10\\-2\\\hline\end{array}$ $\begin{array}{r}5\\-2\\\hline\end{array}$ $\begin{array}{r}11\\-2\\\hline\end{array}$ $\begin{array}{r}6\\-2\\\hline\end{array}$

## 2

**A**

15 { 9 ______ ; [ ] ______

**B**

15 { 8 ______ ; [ ] ______

## 3

$\begin{array}{r}15\\-6\\\hline\end{array}$ $\begin{array}{r}15\\-8\\\hline\end{array}$ $\begin{array}{r}15\\-10\\\hline\end{array}$ $\begin{array}{r}15\\-9\\\hline\end{array}$ $\begin{array}{r}15\\-6\\\hline\end{array}$ $\begin{array}{r}15\\-8\\\hline\end{array}$ $\begin{array}{r}15\\-7\\\hline\end{array}$ $\begin{array}{r}15\\-6\\\hline\end{array}$

$\begin{array}{r}15\\-8\\\hline\end{array}$ $\begin{array}{r}15\\-7\\\hline\end{array}$ $\begin{array}{r}15\\-10\\\hline\end{array}$ $\begin{array}{r}15\\-6\\\hline\end{array}$ $\begin{array}{r}15\\-8\\\hline\end{array}$ $\begin{array}{r}15\\-7\\\hline\end{array}$ $\begin{array}{r}15\\-6\\\hline\end{array}$ $\begin{array}{r}15\\-8\\\hline\end{array}$

## 4

$\begin{array}{r}13\\-5\\\hline\end{array}$ $\begin{array}{r}16\\-9\\\hline\end{array}$ $\begin{array}{r}13\\-9\\\hline\end{array}$ $\begin{array}{r}13\\-6\\\hline\end{array}$ $\begin{array}{r}9\\-3\\\hline\end{array}$ $\begin{array}{r}13\\-7\\\hline\end{array}$ $\begin{array}{r}9\\-5\\\hline\end{array}$ $\begin{array}{r}13\\-4\\\hline\end{array}$

$\begin{array}{r}13\\-9\\\hline\end{array}$ $\begin{array}{r}13\\-8\\\hline\end{array}$ $\begin{array}{r}15\\-9\\\hline\end{array}$ $\begin{array}{r}9\\-6\\\hline\end{array}$ $\begin{array}{r}13\\-4\\\hline\end{array}$ $\begin{array}{r}9\\-4\\\hline\end{array}$ $\begin{array}{r}11\\-9\\\hline\end{array}$ $\begin{array}{r}14\\-6\\\hline\end{array}$

$\begin{array}{r}14\\-9\\\hline\end{array}$ $\begin{array}{r}13\\-5\\\hline\end{array}$ $\begin{array}{r}10\\-3\\\hline\end{array}$ $\begin{array}{r}17\\-9\\\hline\end{array}$ $\begin{array}{r}13\\-8\\\hline\end{array}$ $\begin{array}{r}12\\-9\\\hline\end{array}$ $\begin{array}{r}9\\-7\\\hline\end{array}$ $\begin{array}{r}13\\-5\\\hline\end{array}$

Part 4 continues on the next page.

| 9 | 18 | 13 | 16 | 9 | 13 | 12 | 10 |
|---|---|---|---|---|---|---|---|
| − 4 | − 9 | − 4 | − 9 | − 3 | − 8 | − 5 | − 6 |

**5**

**A** There are 143 cabins up at the lake. This year more cabins were built. Now there are 160. How many more cabins were built at the lake?

[ ] ________________

**B** My dad's store was having a sale. There were 114 people in the store. Some more people came in. Then there were 142 people in the store. How many more people came into the store?

[ ] ________________

**C** Benji had a stamp collection. When the year began, he had 1436 stamps. During the year he got 148 stamps. How many stamps did Benji have at the end of the year?

[ ] ________________

**D** A meal was served at a banquet. There were 143 people sitting at the tables. Some more people arrived. Now there are 160 people at the tables. How many more people arrived?

[ ] ________________

**E** When the game started, there were 842 people watching it. During the game another 58 people began to watch it. How many people were watching the game altogether?

[ ] ________________

## 6

A 800 − 1 = ______ B 600 − 1 = ______ C 400 − 1 = ______

D 100 − 1 = ______ E 700 − 1 = ______ F 200 − 1 = ______

## 7

A Mrs. Kirk owns a car rental business. She rented 46 sports cars. She rented 64 station wagons. How many cars in all did Mrs. Kirk rent?

[ ] ____________________

B 90 children were taking piano lessons. 27 more children started piano lessons. How many children were taking piano lessons then?

[ ] ____________________

C At camp the children hung their bathing suits on clotheslines to dry. This morning there were 32 wet bathing suits on the clothesline. The rest were dry. There were 70 bathing suits in all. How many dry bathing suits were there?

[ ] ____________________

D A farmer delivered 346 boxes of strawberries to a market. After the market opened, 99 boxes of strawberries were sold in one hour. How many boxes of strawberries does the market still have to sell?

[ ] ____________________

Part 7 continues on the next page.

**E** A flower shop sold 27 bunches of daisies in the morning. The shop sold 30 bunches of roses in the afternoon. How many bunches of flowers in all were sold?

[ ] ________________

**F** Mr. Elkins drives a school bus. Today 27 of his passengers were girls. There were 40 passengers in all. How many boys were there?

[ ] ________________

**G** There were 146 cars in a parking lot. There were 94 clean cars. How many dirty cars were in the parking lot?

[ ] ________________

## 8

**A** $\begin{array}{r} 7400 \\ -2739 \\ \hline \end{array}$

**B** $\begin{array}{r} 4100 \\ -\ 328 \\ \hline \end{array}$

**C** $\begin{array}{r} 5134 \\ -\ 994 \\ \hline \end{array}$

**D** $\begin{array}{r} 705 \\ +288 \\ \hline \end{array}$

**E** $\begin{array}{r} 4106 \\ -\ 258 \\ \hline \end{array}$

**F** $\begin{array}{r} 4300 \\ -3662 \\ \hline \end{array}$

**G** $\begin{array}{r} 5052 \\ -4678 \\ \hline \end{array}$

**H** $\begin{array}{r} 3842 \\ +1586 \\ \hline \end{array}$

**I** $\begin{array}{r} 3064 \\ -1998 \\ \hline \end{array}$

**J** $\begin{array}{r} 800 \\ -530 \\ \hline \end{array}$

# Lesson 45

Facts + Problems + Bonus = TOTAL

## 1

| | | | | | | | |
|---|---|---|---|---|---|---|---|
| $7 - 2$ | $11 - 2$ | $5 - 2$ | $8 - 2$ | $9 - 2$ | $6 - 2$ | $10 - 2$ | $7 - 2$ |
| $9 - 2$ | $4 - 2$ | $5 - 2$ | $10 - 2$ | $3 - 2$ | $6 - 2$ | $11 - 2$ | $8 - 2$ |

## 2

**A** 15 { 9, [ ] }

**B** 15 { 8, [ ] }

## 3

| | | | | | | | |
|---|---|---|---|---|---|---|---|
| $15 - 6$ | $15 - 8$ | $15 - 10$ | $15 - 7$ | $15 - 9$ | $15 - 6$ | $15 - 8$ | $15 - 7$ |
| $15 - 9$ | $15 - 6$ | $15 - 7$ | $15 - 8$ | $15 - 6$ | $15 - 7$ | $15 - 9$ | $15 - 8$ |

## 4

| | | | | | | | |
|---|---|---|---|---|---|---|---|
| $13 - 4$ | $9 - 4$ | $13 - 9$ | $13 - 6$ | $16 - 7$ | $14 - 6$ | $13 - 5$ | $9 - 3$ |
| $9 - 6$ | $13 - 6$ | $11 - 9$ | $14 - 8$ | $13 - 4$ | $9 - 5$ | $13 - 8$ | $12 - 9$ |
| $13 - 8$ | $13 - 4$ | $9 - 6$ | $16 - 7$ | $13 - 7$ | $9 - 3$ | $13 - 5$ | $15 - 9$ |

Part 4 continues on the next page.

| 14 | 9 | 16 | 13 | 10 | 9 | 13 | 10 |
|---|---|---|---|---|---|---|---|
| − 9 | − 4 | − 9 | − 7 | − 3 | − 5 | − 9 | − 4 |

**A** When the movie began, there were 146 people in the theater. By the end of the movie there were 206 people in the theater. How many people came into the theater after the movie started?

[ ] ____________________

**B** Last summer we took photographs of a building that was being built. At that time the building had 58 floors. In winter the building was finished. It had 70 floors. How many floors were built after we took photographs?

[ ] ____________________

**C** 315 children took the bus to summer camp. Another 46 children took the train to camp. How many children went to summer camp in all?

[ ] ____________________

**D** At a beach there were 54 swimmers. 40 more swimmers came. How many swimmers were at the beach then?

[ ] ____________________

**E** A dress shop had 23 dresses. Some more dresses came in a truck. Now there are 70 dresses in the shop. How many dresses came in the truck?

[ ] ____________________

A 800 − 1 = ______ B 400 − 1 = ______ C 100 − 1 = ______

D 600 − 1 = ______ E 900 − 1 = ______ F 200 − 1 = ______

## 7

A We had a picnic for some friends. 37 children and 40 adults came. How many people came to the picnic?

[ ] ________________

B The big dinner party at Robert's house is over. He has 42 dirty dishes. There are 60 dishes in all. How many clean dishes does Robert have?

[ ] ________________

C Our family had 230 puzzles. We lost 8 puzzles when we moved. Now how many puzzles do we have left?

[ ] ________________

D Some lumbermen cut down 1456 tall trees in a forest. They cut down 138 short trees. How many trees in all did the lumbermen cut down?

[ ] ________________

E A fruit grower delivered fruit to a hotel kitchen. There were 50 boxes of pears and 24 boxes of plums. How many boxes of fruit did the grower deliver to the hotel kitchen?

[ ] ________________

Part 7 continues on the next page.

**F** There were 240 passengers on an airplane. There were 70 adults. How many were children?

[ ] ____________

**G** Last month our teacher read 90 pages of a mystery story to our class. This month he has read 35 more pages of the story. How many pages has the teacher read?

[ ] ____________

**8**

**A** $6100 - 538$

**B** $300 - 94$

**C** $7052 - 1378$

**D** $945 + 56$

**E** $3100 - 259$

**F** $4300 - 4240$

**G** $8206 - 7949$

**H** $800 - 460$

**I** $78 + 40$

**J** $3928 - 2998$

# Lesson 46

Facts + Problems + Bonus = TOTAL

## 1

$11 - 2$ $7 - 2$ $9 - 2$ $5 - 2$ $4 - 2$ $10 - 2$ $6 - 2$ $8 - 2$

$10 - 2$ $3 - 2$ $7 - 2$ $6 - 2$ $11 - 2$ $8 - 2$ $5 - 2$ $9 - 2$

## 2

$15 - 6$ $15 - 9$ $15 - 7$ $15 - 8$ $15 - 6$ $15 - 10$ $15 - 7$ $15 - 8$

$15 - 7$ $15 - 8$ $15 - 6$ $15 - 9$ $15 - 8$ $15 - 10$ $15 - 6$ $15 - 7$

## 3

$13 - 5$ $11 - 9$ $13 - 4$ $12 - 9$ $13 - 8$ $10 - 4$ $13 - 9$ $13 - 6$

$9 - 4$ $16 - 7$ $14 - 9$ $8 - 3$ $9 - 3$ $13 - 5$ $16 - 7$ $10 - 2$

$9 - 6$ $12 - 4$ $13 - 4$ $9 - 7$ $13 - 9$ $10 - 8$ $13 - 6$ $16 - 9$

$9 - 4$ $15 - 9$ $10 - 7$ $8 - 5$ $17 - 9$ $13 - 4$ $16 - 7$ $12 - 3$

## 4

**A** $800 - 1 =$ ______ **B** $500 - 1 =$ ______ **C** $100 - 1 =$ ______

**D** $600 - 1 =$ ______ **E** $200 - 1 =$ ______ **F** $900 - 1 =$ ______

5

**A** There were 36 nuts in a jar. Some more nuts were put into the jar. Now there are 52 nuts in the jar. How many more nuts were put into the jar?

[ ] ________________

**B** Gail's science class was studying insects. In one week Gail collected 14 insects. The next week she collected another 20 insects. How many did she have then?

[ ] ________________

**C** Gus had learned to play 16 songs on the violin. Then he learned how to play some new songs. Now Gus can play 27 songs. How many new songs did he learn?

[ ] ________________

**D** A school had a spring sports contest. 45 students arrived early. Then later 160 more students arrived. How many students came to the contest?

[ ] ________________

**E** Janet raises rabbits. In March, Janet owned 34 rabbits. In May, she owned 70. How many new rabbits did she get from March to May?

[ ] ________________

6

**A** A waiter served 64 hot drinks to people in a restaurant. He served 90 drinks in all. How many cold drinks did the waiter serve?

[ ] ________________

Part 6 continues on the next page.

**B** During a sale, a men's shop sold 90 leather jackets and 42 wool jackets. How many jackets did the shop sell?

[ ] ____________

**C** On a ranch there are 90 full-grown sheep and 42 lambs. How many animals are on the ranch?

[ ] ____________

**D** There were 90 adult monkeys in a zoo. The zoo had 140 monkeys in all. How many baby monkeys did the zoo have?

[ ] ____________

**E** 530 people work in a factory. On Mondays 80 people leave to go to classes. How many people are left at work?

[ ] ____________

**F** Janet collects coins. On Monday she owned 34 coins. On Friday she owned 70 coins. How many coins did she get from Monday to Friday?

[ ] ____________

**G** We grew roses and daisies in our garden. We picked 3 daisies. We picked 20 flowers in all. How many roses did we pick?

[ ] ____________

**H** In an office there were 940 workers. 80 more workers are hired. How many people work in the office now?

[ ] ____________

## 7

**A**
$$\begin{array}{r} 800 \\ -\ \ 42 \\ \hline \end{array}$$

**B**
$$\begin{array}{r} 7400 \\ -6990 \\ \hline \end{array}$$

**C**
$$\begin{array}{r} 3284 \\ +\ \ 820 \\ \hline \end{array}$$

**D**
$$\begin{array}{r} 900 \\ -\ \ 80 \\ \hline \end{array}$$

**E**
$$\begin{array}{r} 4106 \\ -1328 \\ \hline \end{array}$$

**F**
$$\begin{array}{r} 5300 \\ -\ \ 869 \\ \hline \end{array}$$

**G**
$$\begin{array}{r} 8300 \\ -\ \ 974 \\ \hline \end{array}$$

**H**
$$\begin{array}{r} 3104 \\ -\ \ \ 28 \\ \hline \end{array}$$

**I**
$$\begin{array}{r} 7902 \\ -5896 \\ \hline \end{array}$$

**J**
$$\begin{array}{r} 3842 \\ +1060 \\ \hline \end{array}$$

**K**
$$\begin{array}{r} 5600 \\ +\ \ 699 \\ \hline \end{array}$$

**L**
$$\begin{array}{r} 9012 \\ -8996 \\ \hline \end{array}$$

**M**
$$\begin{array}{r} 5340 \\ -4930 \\ \hline \end{array}$$

**N**
$$\begin{array}{r} 8040 \\ -1036 \\ \hline \end{array}$$

**O**
$$\begin{array}{r} 9012 \\ -8612 \\ \hline \end{array}$$

# Lesson 47

Facts + Problems + Bonus = TOTAL

**1**

| | | | | | | | |
|---|---|---|---|---|---|---|---|
| 11 − 2 | 11 − 5 | 11 − 3 | 11 − 4 | 11 − 3 | 11 − 5 | 11 − 2 | 11 − 5 |

**2**

**A** 15 { 9 ______ ; [ ] ______ }

**B** 15 { 8 ______ ; [ ] ______ }

**3**

| | | | | | | | |
|---|---|---|---|---|---|---|---|
| 15 − 6 | 15 − 10 | 15 − 8 | 15 − 6 | 15 − 9 | 15 − 7 | 15 − 6 | 15 − 9 |
| 15 − 8 | 15 − 6 | 15 − 10 | 15 − 7 | 15 − 9 | 15 − 6 | 15 − 7 | 15 − 8 |

**4**

| | | | | | | | |
|---|---|---|---|---|---|---|---|
| 8 − 2 | 13 − 4 | 13 − 9 | 9 − 2 | 16 − 7 | 13 − 5 | 9 − 4 | 13 − 6 |
| 10 − 4 | 13 − 7 | 9 − 5 | 13 − 8 | 12 − 9 | 6 − 2 | 13 − 6 | 9 − 6 |
| 13 − 5 | 9 − 5 | 11 − 9 | 13 − 9 | 13 − 4 | 13 − 8 | 15 − 9 | 9 − 7 |
| 16 − 7 | 14 − 6 | 13 − 7 | 10 − 6 | 17 − 9 | 10 − 8 | 13 − 5 | 14 − 8 |

**5**

**A** 800 − 1 = ______ **B** 400 − 1 = ______ **C** 100 − 1 = ______

**D** 600 − 1 = ______ **E** 900 − 1 = ______ **F** 300 − 1 = ______

## 6

**A** In a park there is a goldfish pond. 48 goldfish were in the pond. This spring 70 more goldfish were put in the pond. How many goldfish are in the pond now?

[ ] ________________

**B** There was a bad storm last night. 64 trees were smashed in our town. Some more trees were smashed in the next town. 80 trees in all were smashed. How many trees were smashed in the next town?

[ ] ________________

**C** By six o'clock a restaurant had served 14 pizzas. Then it served some more pizzas. By ten o'clock, 22 pizzas had been served. How many pizzas were served from six o'clock until ten o'clock?

[ ] ________________

**D** At the beginning of the year, 14 swimmers were on the team. Later some more swimmers joined the team. At the end of the year, there were 22 swimmers on the team. How many swimmers joined the team during the year?

[ ] ________________

**E** A country fair was held in our town yesterday. 1468 people were there very early. 1680 more people came later. How many people were at the fair then?

[ ] ________________

Part 6 continues on the next page.

**F** Yoko was helping her father build a brick wall. By the beginning of May, they had used 384 bricks. They used some more bricks. By the end of May 830 bricks had been used in building the wall. How many bricks did Yoko and her father use during the month of May?

[ ] ________________

## 7

**A** There is a total of 250 students who go to Grant School. 208 students are present today. How many students are absent?

[ ] ________________

**B** A train was traveling from Toronto to Montreal. There were 145 adults on the train. There were 220 passengers altogether. How many children were on the train?

[ ] ________________

**C** There were 100 clocks in a clock shop. The shopowner bought 84 more clocks. How many clocks are in the shop now?

[ ] ________________

**D** 145 people work in the car factory. 25 more people have been hired to work on Monday. How many people in all will be working on Monday?

[ ] ________________

Part 7 continues on the next page.

**E** Mrs. Franks and Mrs. Schultz sold popcorn. One morning Mrs. Franks sold 20 bags of popcorn. Mrs. Schultz sold 40 bags. How many bags of popcorn did they sell?

[ ] ________________

**F** Jenny has 408 Canadian stamps. She traded 98 stamps for some coins. How many Canadian stamps does Jenny have now?

[ ] ________________

**G** A repair shop fixed 48 watches in the morning. It repaired 90 watches in all that day. How many watches did it repair during the afternoon?

[ ] ________________

**H** A market sold 2350 apples. It sold 2150 oranges. How many pieces of fruit did the market sell?

[ ] ________________

## 8

**A** $\begin{array}{r} 5042 \\ -4980 \\ \hline \end{array}$

**B** $\begin{array}{r} 8700 \\ -7984 \\ \hline \end{array}$

**C** $\begin{array}{r} 3510 \\ -3410 \\ \hline \end{array}$

**D** $\begin{array}{r} 500 \\ -\ 80 \\ \hline \end{array}$

**E** $\begin{array}{r} 7406 \\ -\ 970 \\ \hline \end{array}$

**F** $\begin{array}{r} 908 \\ +\ 38 \\ \hline \end{array}$

**G** $\begin{array}{r} 600 \\ -\ 43 \\ \hline \end{array}$

**H** $\begin{array}{r} 8100 \\ -\ 299 \\ \hline \end{array}$

**I** $\begin{array}{r} 7058 \\ -\ 950 \\ \hline \end{array}$

**J** $\begin{array}{r} 426 \\ -\ 99 \\ \hline \end{array}$

# Lesson 48

| Test | + | Facts | + | Problems | + | Bonus | = | TOTAL |
|---|---|---|---|---|---|---|---|---|
| | | | | | | | | |

## 1

| | | | | | | | |
|---|---|---|---|---|---|---|---|
| 11 | 11 | 11 | 11 | 11 | 11 | 11 | 11 |
| − 2 | − 5 | − 3 | − 4 | − 2 | − 4 | − 3 | − 5 |

## 2

**A** 14 { 9 ________ ; ☐ ________

**B** 17 { 9 ________ ; ☐ ________

**C** 16 { 9 ________ ; ☐ ________

**D** 12 { 9 ________ ; ☐ ________

## 3

| | | | | | | | |
|---|---|---|---|---|---|---|---|
| 15 | 15 | 15 | 15 | 15 | 15 | 15 | 15 |
| − 6 | − 10 | − 7 | − 9 | − 8 | − 6 | − 9 | − 7 |
| 15 | 15 | 15 | 15 | 15 | 15 | 15 | 15 |
| − 10 | − 7 | − 6 | − 9 | − 8 | − 7 | − 6 | − 9 |

## 4

| | | | | | | | |
|---|---|---|---|---|---|---|---|
| 8 | 13 | 7 | 9 | 13 | 8 | 9 | 13 |
| − 2 | − 4 | − 2 | − 5 | − 5 | − 5 | − 2 | − 7 |
| 13 | 10 | 9 | 13 | 12 | 9 | 13 | 14 |
| − 8 | − 2 | − 5 | − 6 | − 9 | − 4 | − 7 | − 9 |
| 13 | 6 | 9 | 13 | 11 | 9 | 13 | 9 |
| − 9 | − 2 | − 4 | − 5 | − 2 | − 6 | − 4 | − 7 |

Part 4 continues on the next page.

| 13 | 16 | 16 | 9 | 13 | 15 | 13 | 17 |
|---|---|---|---|---|---|---|---|
| − 8 | − 9 | − 7 | − 3 | − 4 | − 9 | − 5 | − 9 |

## 5

**A**
$$\begin{array}{r} 6000 \\ -4184 \\ \hline \end{array}$$

**B**
$$\begin{array}{r} 9003 \\ -1248 \\ \hline \end{array}$$

**C**
$$\begin{array}{r} 1004 \\ -\ 368 \\ \hline \end{array}$$

**D**
$$\begin{array}{r} 4000 \\ -\ 82 \\ \hline \end{array}$$

## 6

**A** Gretchen had a cup collection. When the year began, she had 164 cups. At the end of the year, she had 270 cups. How many cups did she get during the year?

[ ] ________________

**B** James had typed 114 pages of a report. Then he typed some more pages. Now he has typed 142 pages. How many more pages did James type?

[ ] ________________

**C** My uncle could play 118 songs on the piano. This year he learned to play 64 more songs. How many songs can my uncle play now?

[ ] ________________

**D** There were 143 ducks swimming in a lake. Some more ducks flew down to the lake. Then there were 160 ducks on the lake. How many ducks flew down to the lake?

[ ] ________________

Part 6 continues on the next page.

**E** Mr. Simmons is a secretary in a large office. On Monday he mailed 34 letters. On Tuesday he mailed 60 letters. How many letters did Mr. Simmons mail on Monday and Tuesday?

[ ] ________________

## 7

**A** Our team lost 42 games. It played 70 games in all. How many games did our team win?

[ ] ________________

**B** A farmer had 136 bunches of carrots. He pulled up 16 more bunches of carrots. How many bunches of carrots did the farmer have then?

[ ] ________________

**C** 140 paintings were on sale at an art show. Some of the paintings were sold. At the end of the show there were 85 paintings left. How many were sold?

[ ] ________________

**D** Mr. Galanos has entered many kinds of contests. He has won prizes in 36 contests. In 42 other contests, he didn't win any prizes. How many contests in all has Mr. Galanos entered?

[ ] ________________

**E** A pet shop sells dog collars and cat collars. It has 64 dog collars. The shop has 80 collars in all. How many cat collars does the shop have?

[ ] ________________

Part 7 continues on the next page.

**F** Margo and Helene are building a table. Margo has used 20 nails so far in building the table. Helene has used 15 nails. How many nails have the girls used?

☐ ____________________

**G** 406 people were at the dance. 9 people went home early. How many people were still at the dance?

☐ ____________________

## 8

**A** $4350 - 4273$

**B** $500 - 86$

**C** $3506 - 719$

**D** $400 - 60$

**E** $4308 - 3530$

**F** $4300 - 926$

**G** $7100 - 824$

**H** $800 - 60$

**I** $3200 - 3184$

**J** $3100 - 1440$

# Lesson 49

Facts + Problems + Bonus = TOTAL

**1**

| 11 | 11 | 11 | 11 | 11 | 11 | 11 | 11 |
|---|---|---|---|---|---|---|---|
| − 4 | − 3 | − 5 | − 2 | − 4 | − 2 | − 4 | − 5 |

**2**

**A** 14 { 9 ______ ; [ ] ______

**B** 17 { 9 ______ ; [ ] ______

**C** 16 { 9 ______ ; [ ] ______

**D** 13 { 9 ______ ; [ ] ______

**3**

| 14 | 15 | 15 | 15 | 14 | 16 | 15 | 13 |
|---|---|---|---|---|---|---|---|
| − 5 | − 7 | − 9 | − 6 | − 5 | − 9 | − 8 | − 4 |

| 14 | 17 | 16 | 13 | 15 | 14 | 15 | 16 |
|---|---|---|---|---|---|---|---|
| − 9 | − 9 | − 7 | − 4 | − 7 | − 5 | − 9 | − 7 |

**4**

| 9 | 15 | 13 | 11 | 15 | 13 | 15 | 13 |
|---|---|---|---|---|---|---|---|
| − 2 | − 6 | − 4 | − 2 | − 8 | − 6 | − 7 | − 8 |

| 13 | 15 | 13 | 7 | 13 | 15 | 17 | 13 |
|---|---|---|---|---|---|---|---|
| − 6 | − 9 | − 5 | − 2 | − 9 | − 6 | − 9 | − 9 |

| 14 | 9 | 15 | 15 | 9 | 11 | 10 | 16 |
|---|---|---|---|---|---|---|---|
| − 9 | − 7 | − 7 | − 8 | − 6 | − 9 | − 2 | − 9 |

Part 4 continues on the next page.

| 12 | 13 | 15 | 13 | 15 | 13 | 13 | 9 |
|---|---|---|---|---|---|---|---|
| − 9 | − 5 | − 6 | − 4 | − 8 | − 7 | − 6 | − 4 |

## 5

| A | B | C | D | E |
|---|---|---|---|---|
| 5000 | 5000 | 5000 | 8000 | 8000 |
| − 3 | − 30 | − 300 | − 4 | − 40 |

| F | G | H | I | J |
|---|---|---|---|---|
| 3000 | 3000 | 1000 | 1000 | 1000 |
| − 8 | − 800 | − 7 | − 20 | − 400 |

## 6

**A** There were 48 horses on the ranch. Some horses were born. Now there are 66 horses. How many horses were born?

____________

**B** This year a shop sold 1463 gray suits. The shop also sold 3652 blue suits. How many suits did the shop sell?

____________

**C** A peach grower sent 400 boxes of peaches to a market. 25 boxes were spoiled when they arrived. How many boxes were all right?

____________

**D** Mrs. Weston collects plates. When she started, she had 134 plates. At the end of a year Mrs. Weston had 200 plates. How many plates did she get during the year?

____________

Part 6 continues on the next page.

**E** Mrs. Carr usually exercises 142 hours every month. Last month she missed some hours. She only exercised for 126 hours. How many hours of exercise did Mrs. Carr miss last month?

[ ] ____________________

**F** Bryan invited 13 children and 20 adults to his mother's birthday party. How many people were invited to the party?

[ ] ____________________

**G** An automobile racing club had 84 members. 14 more members joined the club. How many people belong to the club now?

[ ] ____________________

**H** 143 women went to the town meeting. 280 people in all attended the meeting. How many men attended the meeting?

[ ] ____________________

**I** Mr. Teller owns a school that trains dogs. At the beginning of June 369 dogs had been trained. By the end of June 435 dogs had been trained. How many dogs were trained in June?

[ ] ____________________

## 7

**A** $\begin{array}{r} 6400 \\ -\ \ 986 \\ \hline \end{array}$

**B** $\begin{array}{r} 7100 \\ -\ \ 280 \\ \hline \end{array}$

**C** $\begin{array}{r} 5120 \\ -4370 \\ \hline \end{array}$

**D** $\begin{array}{r} 6012 \\ -4999 \\ \hline \end{array}$

**E** $\begin{array}{r} 4106 \\ -\ \ \ \ 9 \\ \hline \end{array}$

**F** $\begin{array}{r} 5600 \\ +\ \ 699 \\ \hline \end{array}$

**G** $\begin{array}{r} 9012 \\ -8996 \\ \hline \end{array}$

**H** $\begin{array}{r} 5340 \\ -4930 \\ \hline \end{array}$

**I** $\begin{array}{r} 8040 \\ -1036 \\ \hline \end{array}$

**J** $\begin{array}{r} 9012 \\ -8612 \\ \hline \end{array}$

# Lesson 50

| Test | + | Facts | + | Problems | + | Bonus | = | TOTAL |
|---|---|---|---|---|---|---|---|---|
| | | | | | | | | |

## 1

| 11 | 11 | 11 | 11 | 11 | 11 | 11 | 11 |
|---|---|---|---|---|---|---|---|
| − 2 | − 4 | − 3 | − 5 | − 3 | − 2 | − 5 | − 4 |

## 2

**A**

14 { 9 ______ / ☐ ______

**B**

16 { 9 ______ / ☐ ______

## 3

| 14 | 16 | 14 | 14 | 16 | 16 | 13 | 14 |
|---|---|---|---|---|---|---|---|
| − 9 | − 7 | − 5 | − 8 | − 7 | − 8 | − 8 | − 5 |

| 13 | 16 | 13 | 14 | 13 | 16 | 13 | 14 |
|---|---|---|---|---|---|---|---|
| − 4 | − 7 | − 9 | − 6 | − 4 | − 8 | − 9 | − 5 |

## 4

| 15 | 13 | 8 | 13 | 15 | 9 | 11 | 15 |
|---|---|---|---|---|---|---|---|
| − 6 | − 8 | − 2 | − 6 | − 7 | − 2 | − 2 | − 8 |

| 13 | 9 | 13 | 16 | 15 | 7 | 13 | 15 |
|---|---|---|---|---|---|---|---|
| − 7 | − 2 | − 5 | − 8 | − 8 | − 2 | − 7 | − 9 |

| 10 | 6 | 15 | 13 | 16 | 15 | 13 | 5 |
|---|---|---|---|---|---|---|---|
| − 4 | − 2 | − 8 | − 4 | − 8 | − 7 | − 6 | − 4 |

| 13 | 10 | 15 | 12 | 4 | 13 | 12 | 13 |
|---|---|---|---|---|---|---|---|
| − 8 | − 3 | − 6 | − 5 | − 2 | − 5 | − 8 | − 4 |

## 5

**A** 90 children went to a play. 46 of the children enjoyed the play. The rest did not. How many of the children did not enjoy the play?

**B** We bought some eggs. When we got home, 36 eggs were broken. 24 eggs were not broken. How many eggs in all did we buy?

**C** There are 62 sharks. 45 of the sharks are hungry. How many of the sharks are not hungry?

**D** 80 people were watching a bulldozer knock down a building. 69 people left to return to work. How many people stayed to watch?

**E** Linda Arrowhead's father is a baker. One Saturday Linda helped her father make pies for some restaurants. They made 60 apple pies and 45 pies that were not apple. How many pies in all did Linda and her father make?

## 6

**A** $3000 - 8$

**B** $3000 - 800$

**C** $3000 - 80$

**D** $1000 - 20$

**E** $1000 - 2$

**F** $4000 - 245$

**G** $4000 - 2450$

**H** $8600 - 342$

**I** $8000 - 300$

**J** $6000 - 4128$

## 7

**A** 136 people had paid for swimming lessons. Then some more people paid for swimming lessons. Now 180 people have paid for lessons. How many more people paid for swimming lessons?

[ ] __________________

**B** A greeting card shop sold 43 cards in the morning. It sold 50 cards in the afternoon. How many cards in all did it sell?

[ ] __________________

**C** There were 3170 cars in the parking garage. There were 990 dirty cars. How many clean cars were there?

[ ] __________________

**D** When we first moved to Morris, 2683 people lived here. Now 3780 people live in Morris. How many people have moved to Morris?

[ ] __________________

**E** Carmen ran for 37 hours this week. Elena ran for 40 hours this week. How many hours in all did the girls run this week?

[ ] __________________

**F** Reggie had 70 old English coins. He gave some of them to his brother. Now Reggie has 48 old English coins. How many coins did he give to his brother?

[ ] __________________

Part 7 continues on the next page.

**G** Mr. Yamada visits a used-book shop every week. Mr. Yamada had 36 books. Last week he bought some more books at the used-book shop. Now he has 58 books. How many books did he buy last week?

[ ] ____________________

**H** An office building has 2365 clean windows. The window washers have to wash 90 dirty windows. How many windows in all does the building have?

[ ] ____________________

## 8

**A** $\begin{array}{r} 3104 \\ -\ \ 78 \\ \hline \end{array}$

**B** $\begin{array}{r} 4106 \\ -\ 256 \\ \hline \end{array}$

**C** $\begin{array}{r} 3460 \\ -1990 \\ \hline \end{array}$

**D** $\begin{array}{r} 5360 \\ -5298 \\ \hline \end{array}$

**E** $\begin{array}{r} 4300 \\ -3996 \\ \hline \end{array}$

**F** $\begin{array}{r} 5300 \\ -\ 869 \\ \hline \end{array}$

**G** $\begin{array}{r} 8300 \\ -\ 974 \\ \hline \end{array}$

**H** $\begin{array}{r} 3104 \\ -\ \ 28 \\ \hline \end{array}$

**I** $\begin{array}{r} 7902 \\ -5896 \\ \hline \end{array}$

**J** $\begin{array}{r} 7400 \\ -6990 \\ \hline \end{array}$

# Lesson 51

| Facts | + | Problems | + | Bonus | = | TOTAL |
|---|---|---|---|---|---|---|
| | | | | | | |

## 1

$\begin{array}{r} 11 \\ -\ 2 \\ \hline \end{array}$ $\begin{array}{r} 11 \\ -\ 3 \\ \hline \end{array}$ $\begin{array}{r} 11 \\ -\ 5 \\ \hline \end{array}$ $\begin{array}{r} 11 \\ -\ 4 \\ \hline \end{array}$ $\begin{array}{r} 11 \\ -\ 3 \\ \hline \end{array}$ $\begin{array}{r} 11 \\ -\ 5 \\ \hline \end{array}$ $\begin{array}{r} 11 \\ -\ 4 \\ \hline \end{array}$ $\begin{array}{r} 11 \\ -\ 2 \\ \hline \end{array}$

## 2

**A**

$17 \begin{cases} 9 & \text{\_\_\_\_\_\_\_\_} \\ \square & \text{\_\_\_\_\_\_\_\_} \end{cases}$

**B**

$14 \begin{cases} 9 & \text{\_\_\_\_\_\_\_\_} \\ \square & \text{\_\_\_\_\_\_\_\_} \end{cases}$

## 3

$\begin{array}{r} 17 \\ -\ 9 \\ \hline \end{array}$ $\begin{array}{r} 14 \\ -\ 6 \\ \hline \end{array}$ $\begin{array}{r} 17 \\ -\ 8 \\ \hline \end{array}$ $\begin{array}{r} 13 \\ -\ 6 \\ \hline \end{array}$ $\begin{array}{r} 14 \\ -\ 5 \\ \hline \end{array}$ $\begin{array}{r} 18 \\ -\ 9 \\ \hline \end{array}$ $\begin{array}{r} 16 \\ -\ 7 \\ \hline \end{array}$ $\begin{array}{r} 12 \\ -\ 9 \\ \hline \end{array}$

$\begin{array}{r} 14 \\ -\ 5 \\ \hline \end{array}$ $\begin{array}{r} 11 \\ -\ 9 \\ \hline \end{array}$ $\begin{array}{r} 14 \\ -\ 8 \\ \hline \end{array}$ $\begin{array}{r} 13 \\ -\ 4 \\ \hline \end{array}$ $\begin{array}{r} 15 \\ -\ 9 \\ \hline \end{array}$ $\begin{array}{r} 17 \\ -\ 8 \\ \hline \end{array}$ $\begin{array}{r} 15 \\ -\ 9 \\ \hline \end{array}$ $\begin{array}{r} 16 \\ -\ 7 \\ \hline \end{array}$

## 4

$\begin{array}{r} 15 \\ -\ 6 \\ \hline \end{array}$ $\begin{array}{r} 11 \\ -\ 2 \\ \hline \end{array}$ $\begin{array}{r} 13 \\ -\ 8 \\ \hline \end{array}$ $\begin{array}{r} 10 \\ -\ 2 \\ \hline \end{array}$ $\begin{array}{r} 15 \\ -\ 7 \\ \hline \end{array}$ $\begin{array}{r} 13 \\ -\ 6 \\ \hline \end{array}$ $\begin{array}{r} 8 \\ -2 \\ \hline \end{array}$ $\begin{array}{r} 15 \\ -\ 8 \\ \hline \end{array}$

$\begin{array}{r} 14 \\ -\ 9 \\ \hline \end{array}$ $\begin{array}{r} 15 \\ -\ 7 \\ \hline \end{array}$ $\begin{array}{r} 13 \\ -\ 9 \\ \hline \end{array}$ $\begin{array}{r} 9 \\ -2 \\ \hline \end{array}$ $\begin{array}{r} 13 \\ -\ 6 \\ \hline \end{array}$ $\begin{array}{r} 11 \\ -\ 9 \\ \hline \end{array}$ $\begin{array}{r} 15 \\ -\ 8 \\ \hline \end{array}$ $\begin{array}{r} 9 \\ -2 \\ \hline \end{array}$

$\begin{array}{r} 13 \\ -\ 5 \\ \hline \end{array}$ $\begin{array}{r} 6 \\ -2 \\ \hline \end{array}$ $\begin{array}{r} 13 \\ -\ 8 \\ \hline \end{array}$ $\begin{array}{r} 15 \\ -\ 7 \\ \hline \end{array}$ $\begin{array}{r} 12 \\ -\ 9 \\ \hline \end{array}$ $\begin{array}{r} 13 \\ -\ 7 \\ \hline \end{array}$ $\begin{array}{r} 5 \\ -2 \\ \hline \end{array}$ $\begin{array}{r} 13 \\ -\ 5 \\ \hline \end{array}$

$\begin{array}{r} 13 \\ -\ 6 \\ \hline \end{array}$ $\begin{array}{r} 9 \\ -4 \\ \hline \end{array}$ $\begin{array}{r} 13 \\ -\ 8 \\ \hline \end{array}$ $\begin{array}{r} 16 \\ -\ 9 \\ \hline \end{array}$ $\begin{array}{r} 15 \\ -\ 6 \\ \hline \end{array}$ $\begin{array}{r} 9 \\ -3 \\ \hline \end{array}$ $\begin{array}{r} 13 \\ -\ 5 \\ \hline \end{array}$ $\begin{array}{r} 9 \\ -6 \\ \hline \end{array}$

**A** My sister made a beaded belt. She used 140 large beads to make the belt. 90 of the beads are red. How many of the beads are not red?

**B** There's a new statue in our city. Lots of people saw the statue. 80 people liked the statue. 57 people did not like the statue. How many people saw the statue?

**C** There are 140 people on the beach. 95 of them are wearing hats. How many of the people are not wearing hats?

**D** There were 86 earthquakes last year. 29 of the earthquakes were in Europe. How many earthquakes were not in Europe?

**E** We made sandwiches for our club's picnic. The people who came ate 95 ham sandwiches and 120 chicken sandwiches. How many sandwiches were eaten at the picnic?

**6**

**A** $\begin{array}{r} 8000 \\ -\quad 3 \\ \hline \end{array}$

**B** $\begin{array}{r} 8000 \\ -\quad 30 \\ \hline \end{array}$

**C** $\begin{array}{r} 4000 \\ -\quad 352 \\ \hline \end{array}$

**D** $\begin{array}{r} 4000 \\ -3500 \\ \hline \end{array}$

**E** $\begin{array}{r} 1000 \\ -\quad 80 \\ \hline \end{array}$

**F** $\begin{array}{r} 1000 \\ -\quad 8 \\ \hline \end{array}$

**G** $\begin{array}{r} 7000 \\ -1430 \\ \hline \end{array}$

**H** $\begin{array}{r} 5000 \\ -1008 \\ \hline \end{array}$

**I** $\begin{array}{r} 3000 \\ -\quad 125 \\ \hline \end{array}$

**J** $\begin{array}{r} 1000 \\ -\quad 946 \\ \hline \end{array}$

## 7

**A** Our family is very large. We have 14 first cousins and 20 second cousins. How many cousins are there in our family?

[ ] ____________________

**B** The racing team won 47 car races. They entered 52 car races. How many races did the team lose?

[ ] ____________________

**C** At our club's party there were 42 children and 70 adults. How many people came to the party?

[ ] ____________________

**D** Linda is collecting rocks. This morning she had 136 rocks. Tonight she has 142 rocks. How many rocks did she get during the day?

[ ] ____________________

**E** Benny bought 420 file folders. 254 of the folders have been used. How many are new?

[ ] ____________________

**F** Sergio worked 36 crossword puzzles. Adolfo worked 40 crossword puzzles. How many puzzles did the boys work in all?

[ ] ____________________

Part 7 continues on the next page.

**G** Our stamp club had 126 members last year. Some more people joined the club. Now there are 142 members in the club. How many people joined the club?

[ ] ___________________

**H** A friend of mine has read 462 mystery books and some sports books. She has read 500 books in all. How many sports books has she read?

[ ] ___________________

**I** Don counted 24 sailboats on the lake yesterday. Today Don counted 30 sailboats on the lake. How many boats did Don see on the lake?

[ ] ___________________

## 8

**A**
$$\begin{array}{r} 5100 \\ -\quad 86 \\ \hline \end{array}$$

**B**
$$\begin{array}{r} 3200 \\ -2164 \\ \hline \end{array}$$

**C**
$$\begin{array}{r} 8100 \\ -\quad 364 \\ \hline \end{array}$$

**D**
$$\begin{array}{r} 4600 \\ -3920 \\ \hline \end{array}$$

**E**
$$\begin{array}{r} 5400 \\ -1923 \\ \hline \end{array}$$

# Lesson 52

Facts + Problems + Bonus = TOTAL

## 1

| 17 | 14 | 17 | 13 | 17 | 16 | 15 | 13 |
|---|---|---|---|---|---|---|---|
| − 8 | − 7 | − 9 | − 9 | − 8 | − 8 | − 9 | − 4 |

| 15 | 12 | 14 | 17 | 13 | 14 | 16 | 14 |
|---|---|---|---|---|---|---|---|
| − 6 | − 9 | − 9 | − 8 | − 7 | − 9 | − 7 | − 5 |

## 2

**A** 11 { 4 ______ ; [ ] ______

**B** 11 { 2 ______ ; [ ] ______

**C** 11 { 5 ______ ; [ ] ______

**D** 11 { 3 ______ ; [ ] ______

## 3

| 11 | 11 | 11 | 11 | 11 | 11 | 11 | 11 |
|---|---|---|---|---|---|---|---|
| − 6 | − 8 | − 5 | − 7 | − 8 | − 6 | − 4 | − 7 |

| 11 | 11 | 11 | 11 | 11 | 11 | 11 | 11 |
|---|---|---|---|---|---|---|---|
| − 7 | − 9 | − 6 | − 8 | − 7 | − 2 | − 6 | − 3 |

## 4

| 17 | 13 | 8 | 15 | 13 | 11 | 6 | 15 |
|---|---|---|---|---|---|---|---|
| − 8 | − 5 | − 2 | − 6 | − 6 | − 2 | − 2 | − 8 |

| 16 | 15 | 11 | 10 | 17 | 12 | 14 | 13 |
|---|---|---|---|---|---|---|---|
| − 9 | − 7 | − 9 | − 4 | − 9 | − 8 | − 9 | − 5 |

Part 4 continues on the next page.

| | | | | | | | |
|---|---|---|---|---|---|---|---|
| 15 | 9 | 12 | 13 | 13 | 15 | 13 | 5 |
| − 9 | − 2 | − 3 | − 7 | − 6 | − 7 | − 9 | − 2 |

| | | | | | | | |
|---|---|---|---|---|---|---|---|
| 17 | 13 | 12 | 7 | 15 | 13 | 15 | 16 |
| − 9 | − 8 | − 9 | − 2 | − 9 | − 6 | − 8 | − 9 |

## 5

**A** Melanie did 40 subtraction problems. 17 of them are wrong. How many of Melanie's problems are not wrong?

[ ] ____________________

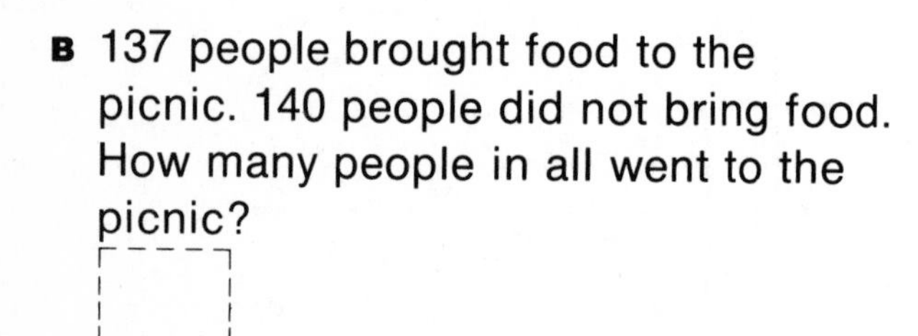

**B** 137 people brought food to the picnic. 140 people did not bring food. How many people in all went to the picnic?

[ ] ____________________

**C** 300 runners entered a race. 27 runners left the race. How many runners did not leave the race?

[ ] ____________________

**D** There were 68 people in front of city hall. 14 of them were police officers. How many people were not police officers?

[ ] ____________________

**E** On Monday 48 people in a large office came to work on time. 7 people did not come to work on time. How many people in all worked in the office on Monday?

[ ] ____________________

## 6

**A** $\begin{array}{r} 8000 \\ -\quad 4 \\ \hline \end{array}$

**B** $\begin{array}{r} 9000 \\ -\quad 30 \\ \hline \end{array}$

**C** $\begin{array}{r} 6000 \\ -\quad 235 \\ \hline \end{array}$

**D** $\begin{array}{r} 1000 \\ -\quad 800 \\ \hline \end{array}$

**E** $\begin{array}{r} 1000 \\ -\quad 927 \\ \hline \end{array}$

**F** $\begin{array}{r} 4000 \\ -\quad 120 \\ \hline \end{array}$

**G** $\begin{array}{r} 6000 \\ -\quad 305 \\ \hline \end{array}$

**H** $\begin{array}{r} 2000 \\ -\quad 125 \\ \hline \end{array}$

**I** $\begin{array}{r} 4000 \\ -\quad 300 \\ \hline \end{array}$

**J** $\begin{array}{r} 1000 \\ -\quad 120 \\ \hline \end{array}$

## 7

**A** A bicycle factory made 4380 bicycles. Then they made another 1560 bicycles. How many bicycles did the factory make?

[ ] ____________________

**B** 140 people went fishing at a large lake. 65 of them caught fish. How many people did not catch any fish?

[ ] ____________________

**C** There were 199 girls inside the school. Some more girls entered the school. Now there are 215 girls inside the school. How many more girls entered the school?

[ ] ____________________

**D** At a band concert there are 1824 men and 2400 women. How many people are at the concert?

[ ] ____________________

Part 7 continues on the next page.

**E** Our team won 34 games. But the team also lost 30 games. How many games did the team play?

[ ] ____________________

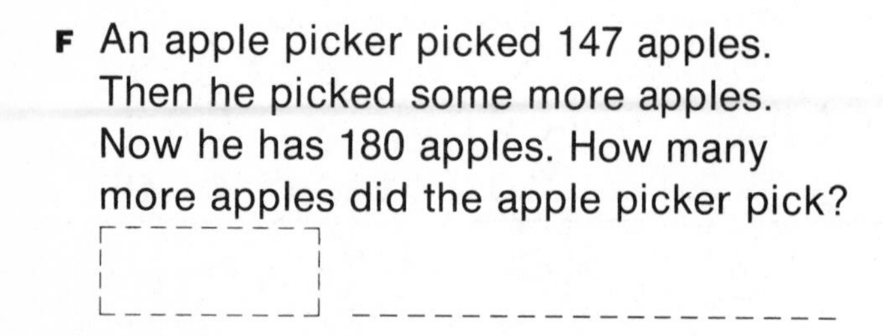

**F** An apple picker picked 147 apples. Then he picked some more apples. Now he has 180 apples. How many more apples did the apple picker pick?

[ ] ____________________

**G** 1460 girls were in a parade. 2320 children were in the parade. How many boys were in the parade?

[ ] ____________________

**H** On a cattle ranch there are 458 cows. There are also 280 bulls on the ranch. How many animals are there on the ranch?

[ ] ____________________

## 8

**A** $\begin{array}{r} 6104 \\ -\quad 29 \\ \hline \end{array}$

**B** $\begin{array}{r} 8107 \\ -\quad 352 \\ \hline \end{array}$

**C** $\begin{array}{r} 210 \\ -\quad 74 \\ \hline \end{array}$

**D** $\begin{array}{r} 846 \\ +352 \\ \hline \end{array}$

**E** $\begin{array}{r} 9062 \\ -8980 \\ \hline \end{array}$

# Lesson 53

| Test | + | Facts | + | Problems | + | Bonus | = | TOTAL |
|---|---|---|---|---|---|---|---|---|
| | | | | | | | | |

## 1

$11 - 8$   $11 - 2$   $11 - 5$   $11 - 3$   $11 - 4$   $11 - 7$   $11 - 9$   $11 - 6$

$11 - 5$   $11 - 7$   $11 - 6$   $11 - 4$   $11 - 8$   $11 - 3$   $11 - 9$   $11 - 7$

## 2

**A** 16 { 9 ____ ; ☐ ____

**B** 13 { 9 ____ ; ☐ ____

**C** 15 { 9 ____ ; ☐ ____

**D** 17 { 9 ____ ; ☐ ____

## 3

$13 - 4$   $17 - 9$   $15 - 8$   $15 - 6$   $13 - 8$   $16 - 7$   $12 - 9$   $14 - 5$

$15 - 9$   $14 - 8$   $16 - 7$   $13 - 9$   $18 - 9$   $14 - 5$   $17 - 9$   $13 - 4$

## 4

$13 - 9$   $15 - 7$   $13 - 6$   $9 - 4$   $15 - 8$   $13 - 7$   $15 - 6$   $15 - 7$

$10 - 4$   $15 - 7$   $15 - 8$   $13 - 7$   $13 - 5$   $15 - 8$   $13 - 6$   $9 - 5$

Part 4 continues on the next page.

| 13 | 9 | 15 | 8 | 13 | 9 | 8 | 15 |
|---|---|---|---|---|---|---|---|
| − 5 | − 6 | − 7 | − 2 | − 8 | − 6 | − 2 | − 8 |

| 10 | 15 | 16 | 8 | 10 | 13 | 10 | 8 |
|---|---|---|---|---|---|---|---|
| − 4 | − 9 | − 9 | − 3 | − 7 | − 9 | − 2 | − 5 |

## 5

**A** There were 140 people on the train. 97 of them were reading a newspaper. How many people were not reading a newspaper?

[ ] ________________

**B** A reporter talked to some people at an art show. 48 people said they liked the art show. 72 people said they did not like the art show. How many people did the reporter talk to?

[ ] ________________

**C** Mrs. Kiyo planted 18 flower seeds. She also planted 40 seeds that were not flower seeds. How many seeds did Mrs. Kiyo plant?

[ ] ________________

**D** Natasha counted 94 cars. 87 of them were new cars. How many were not new cars?

[ ] ________________

**E** There are 150 students. 79 students entered the spelling contest. How many students did not enter the spelling contest?

[ ] ________________

## 6

A
$$\begin{array}{r} 6000 \\ -\ 300 \\ \hline \end{array}$$

B
$$\begin{array}{r} 4000 \\ -1284 \\ \hline \end{array}$$

C
$$\begin{array}{r} 3000 \\ -\ 18 \\ \hline \end{array}$$

D
$$\begin{array}{r} 1000 \\ -\ 20 \\ \hline \end{array}$$

E
$$\begin{array}{r} 6000 \\ -\ 150 \\ \hline \end{array}$$

F
$$\begin{array}{r} 1000 \\ -\ 246 \\ \hline \end{array}$$

G
$$\begin{array}{r} 3000 \\ -\ 400 \\ \hline \end{array}$$

H
$$\begin{array}{r} 8000 \\ -3460 \\ \hline \end{array}$$

I
$$\begin{array}{r} 6000 \\ -\ 145 \\ \hline \end{array}$$

J
$$\begin{array}{r} 9000 \\ -8820 \\ \hline \end{array}$$

## 7

**A** Felipe collected bottles. He had 564 bottles. He found some more bottles. Now he has 610 bottles. How many more bottles did he find?

[ ] ________________

**B** Our principal, Miss DiBono, gave prizes to 14 boys. She gave prizes to 28 girls. How many children got prizes?

[ ] ________________

**C** Mr. Collins is a bookseller. He sold 420 sports books. He also sold 680 mystery books. How many books did he sell?

[ ] ________________

**D** A zoo had 385 birds that sang. The zoo also had 37 other birds that talked. How many birds in all did the zoo have?

[ ] ________________

Part 7 continues on the next page.

**E** The store bought 1640 white boxes. It also bought some brown boxes. It bought 2100 boxes in all. How many brown boxes did the store buy?

**F** 143 shops are open on Sunday. 583 shops are not open. How many shops are there in all?

**G** A flower shop had a big sale. It sold 436 flowers on the first day. By the end of the second day it had sold 544 flowers in all. How many flowers were sold on the second day?

**H** An office building is being built near our school. Workers finished 94 floors of the building last month. This month they finished 19 more floors. Now the building is ready. How many floors does it have?

# Lesson 54

| Test | + | Facts | + | Problems | + | Bonus | = | TOTAL |
|---|---|---|---|---|---|---|---|---|
| | | | | | | | | |

## 1

| 11 | 11 | 11 | 11 | 11 | 11 | 11 | 11 |
|---|---|---|---|---|---|---|---|
| − 8 | − 2 | − 7 | − 4 | − 6 | − 9 | − 3 | − 7 |

| 11 | 11 | 11 | 11 | 11 | 11 | 11 | 11 |
|---|---|---|---|---|---|---|---|
| − 6 | − 4 | − 8 | − 6 | − 5 | − 7 | − 3 | − 8 |

## 2

**A**

17 { 9 ________ / [ ] ________

**B**

14 { 9 ________ / [ ] ________

**C**

15 { 9 ________ / [ ] ________

**D**

13 { 9 ________ / [ ] ________

## 3

| 17 | 13 | 14 | 17 | 16 | 17 | 14 | 13 |
|---|---|---|---|---|---|---|---|
| − 8 | − 9 | − 5 | − 9 | − 9 | − 8 | − 7 | − 9 |

| 14 | 15 | 14 | 15 | 17 | 15 | 17 | 15 |
|---|---|---|---|---|---|---|---|
| − 5 | − 8 | − 9 | − 6 | − 9 | − 8 | − 9 | − 6 |

## 4

| 15 | 13 | 16 | 8 | 15 | 15 | 13 | 10 |
|---|---|---|---|---|---|---|---|
| − 8 | − 6 | − 7 | − 2 | − 6 | − 7 | − 8 | − 2 |

| 9 | 15 | 15 | 7 | 13 | 13 | 13 | 15 |
|---|---|---|---|---|---|---|---|
| − 2 | − 7 | − 8 | − 2 | − 6 | − 7 | − 5 | − 7 |

Part 4 continues on the next page.

| 13 | 6 | 13 | 13 | 9 | 15 | 13 | 12 |
|---|---|---|---|---|---|---|---|
| − 8 | − 4 | − 7 | − 4 | − 4 | − 8 | − 6 | − 3 |

| 13 | 15 | 9 | 15 | 13 | 10 | 10 | 15 |
|---|---|---|---|---|---|---|---|
| − 8 | − 7 | − 4 | − 8 | − 5 | − 6 | − 3 | − 8 |

## 5

**A** There were 62 people at the game. 45 people cheered for the blue team. How many people did not cheer for the blue team?

[ ] ____________________

**B** 58 people on a boat were sick. 70 people were not sick. How many people were on the boat?

[ ] ____________________

**C** 82 people are working on a ship. Today only 47 people came to work. How many people did not come to work?

[ ] ____________________

**D** Some tree cutters cut 90 trees in a forest. They found bugs in 27 of the trees. How many trees did not have bugs?

[ ] ____________________

**E** We live in a large house with many windows. Today 25 of the windows are open. 15 windows are not open. How many windows does our house have?

[ ] ____________________

6

**A** A factory has made 5360 machines. The factory has sold 4890 machines. How many machines does the factory have left to sell?

[ ] ________________

**B** When a large market opened, it gave away balloons. The market gave away 86 red balloons and 90 green balloons. How many balloons did the market give away?

[ ] ________________

**C** 98 people work in a big office. 18 of these people are men. How many of these people are women?

[ ] ________________

**D** There were 1485 liters of water in the swimming pool. More water was added to the pool. Now there are 2095 liters of water in the pool. How many liters of water were added?

[ ] ________________

**E** On Sunday there were 148 fishing boats on the lake. There were also 473 sailboats on the lake. How many boats were on the lake?

[ ] ________________

Part 6 continues on the next page.

**F** Mrs. Ito runs a newsstand in front of a big office building. In the morning she sold 138 newspapers. She sold some more papers in the afternoon. By the end of the day she had sold 150 newspapers. How many newspapers did she sell in the afternoon?

[ ] ____________________

**G** Mr. Campos is a guide in an art museum. Yesterday he was preparing to take 85 people through the museum. At the last moment 17 more people joined the group. How many people did Mr. Campos take through the museum?

[ ] ____________________

## 7

| A | B | C | D | E |
|---|---|---|---|---|
| $5000 - 1402$ | $3000 - 140$ | $1000 - 230$ | $1000 - 156$ | $6000 - 200$ |

| F | G | H | I | J |
|---|---|---|---|---|
| $9004 - 136$ | $3006 - 106$ | $4000 - 125$ | $1004 - 83$ | $5003 - 3999$ |

| K | L | M | N | O |
|---|---|---|---|---|
| $4102 - 86$ | $5103 - 73$ | $8210 - 7946$ | $920 - 814$ | $7100 - 386$ |

# Lesson 55

Facts + Problems + Bonus = TOTAL

## 1

| 11 | 11 | 11 | 11 | 11 | 11 | 11 | 11 |
|---|---|---|---|---|---|---|---|
| − 7 | − 4 | − 5 | − 8 | − 3 | − 7 | − 6 | − 9 |

| 11 | 11 | 11 | 11 | 11 | 11 | 11 | 11 |
|---|---|---|---|---|---|---|---|
| − 4 | − 6 | − 2 | − 3 | − 8 | − 2 | − 5 | − 7 |

## 2

**A** 13 { 9 ______ ; [ ] ______

**B** 16 { 9 ______ ; [ ] ______

**C** 17 { 9 ______ ; [ ] ______

**D** 15 { 9 ______ ; [ ] ______

## 3

| 12 | 15 | 13 | 14 | 17 | 14 | 15 | 15 |
|---|---|---|---|---|---|---|---|
| − 9 | − 6 | − 9 | − 7 | − 8 | − 9 | − 6 | − 8 |

| 16 | 13 | 13 | 17 | 16 | 14 | 13 | 13 |
|---|---|---|---|---|---|---|---|
| − 7 | − 9 | − 5 | − 8 | − 9 | − 5 | − 8 | − 4 |

## 4

| 13 | 13 | 9 | 13 | 15 | 13 | 15 | 13 |
|---|---|---|---|---|---|---|---|
| − 5 | − 7 | − 4 | − 4 | − 7 | − 8 | − 8 | − 6 |

| 9 | 13 | 9 | 13 | 16 | 13 | 9 | 16 |
|---|---|---|---|---|---|---|---|
| − 5 | − 7 | − 2 | − 5 | − 9 | − 8 | − 6 | − 7 |

Part 4 continues on the next page.

$$\begin{array}{r}13\\-\;5\\\hline\end{array}\quad\begin{array}{r}15\\-\;8\\\hline\end{array}\quad\begin{array}{r}8\\-2\\\hline\end{array}\quad\begin{array}{r}10\\-\;6\\\hline\end{array}\quad\begin{array}{r}9\\-3\\\hline\end{array}\quad\begin{array}{r}13\\-\;8\\\hline\end{array}\quad\begin{array}{r}10\\-\;4\\\hline\end{array}\quad\begin{array}{r}15\\-\;7\\\hline\end{array}$$

$$\begin{array}{r}13\\-\;8\\\hline\end{array}\quad\begin{array}{r}15\\-\;7\\\hline\end{array}\quad\begin{array}{r}8\\-3\\\hline\end{array}\quad\begin{array}{r}10\\-\;6\\\hline\end{array}\quad\begin{array}{r}8\\-5\\\hline\end{array}\quad\begin{array}{r}10\\-\;4\\\hline\end{array}\quad\begin{array}{r}10\\-\;2\\\hline\end{array}\quad\begin{array}{r}14\\-\;6\\\hline\end{array}$$

## 5

**A** Mrs. Savas is 42 years old. Miss Hark is 70 years old. How many years older is Miss Hark?

**B** After a few games of table tennis, Amos Silverheels had scored 26 points. His brother Mike had scored 40 points. How many points did Amos and Mike score in all?

**C** Gina weighs 48 kilograms. Sam weighs 29 kilograms. How many more kilograms does Gina weigh?

**D** Mel exercised for 48 minutes. Brad exercised for 42 minutes. How many minutes did Mel and Brad exercise in all?

**E** A building is 36 meters tall. A tree near the building is 48 meters tall. How many meters taller is the tree?

## 6

**A** Swimmers found 1500 coins underwater. 896 of the coins were gold. How many coins were not gold?

[ ] ________________

Part 6 continues on the next page.

**B** 1536 people were at the boat races. Then some more people arrived. Now there are 1600 people at the boat races. How many more people arrived?

☐ ______________

**C** 1504 people took driving tests. 1486 people passed their driving tests. How many people did not pass their driving tests?

☐ ______________

**D** Dan had 48 Canadian stamps. His cousin sent him 26 Canadian stamps. How many Canadian stamps does Dan have in all?

☐ ______________

**E** Jennie jumped rope 140 times last week. Lisa jumped rope 362 times. How many times did Jennie and Lisa jump rope?

☐ ______________

**F** Miss Benton answered 945 telephone calls on Monday. On Tuesday she answered some more telephone calls. Now she has answered 1200 telephone calls. How many telephone calls did Miss Benton answer on Tuesday?

☐ ______________

**G** 195 students are in classes at Oak School. 24 students are absent from classes. How many students in all go to Oak School?

☐ ______________

Part 6 continues on the next page.

**H** The builders had 1530 nails. They used up most of the nails. Now they have 86 nails left. How many nails did they use?

[ ] ____________

**I** Last week we counted the cars that drove past our house. 3500 cars went by. 2850 cars followed the speed limit. How many cars did not follow the speed limit?

[ ] ____________

**J** At one summer camp there were 1368 girls. There are 2100 children at the camp. How many of the children are boys?

[ ] ____________

## 7

**A** $\begin{array}{r} 5000 \\ -\ 324 \\ \hline \end{array}$

**B** $\begin{array}{r} 1000 \\ -\ 230 \\ \hline \end{array}$

**C** $\begin{array}{r} 7204 \\ -\ 264 \\ \hline \end{array}$

**D** $\begin{array}{r} 3100 \\ -2134 \\ \hline \end{array}$

**E** $\begin{array}{r} 1000 \\ -\ 364 \\ \hline \end{array}$

**F** $\begin{array}{r} 9200 \\ -\ 235 \\ \hline \end{array}$

**G** $\begin{array}{r} 6000 \\ -4200 \\ \hline \end{array}$

**H** $\begin{array}{r} 8000 \\ -7940 \\ \hline \end{array}$

**I** $\begin{array}{r} 1000 \\ -\ 429 \\ \hline \end{array}$

**J** $\begin{array}{r} 3005 \\ -2204 \\ \hline \end{array}$

**K** $\begin{array}{r} 6024 \\ -\ 188 \\ \hline \end{array}$

**L** $\begin{array}{r} 5100 \\ -\ 480 \\ \hline \end{array}$

**M** $\begin{array}{r} 7024 \\ -3614 \\ \hline \end{array}$

**N** $\begin{array}{r} 8326 \\ -1996 \\ \hline \end{array}$

**O** $\begin{array}{r} 5034 \\ -4628 \\ \hline \end{array}$

# Lesson 56

| Facts | + | Problems | + | Bonus | = | TOTAL |
|---|---|---|---|---|---|---|
| | | | | | | |

## 1

| | | | | | | | |
|---|---|---|---|---|---|---|---|
| 11 − 6 | 11 − 4 | 11 − 8 | 11 − 9 | 11 − 6 | 11 − 3 | 11 − 8 | 11 − 7 |
| 11 − 5 | 11 − 7 | 11 − 3 | 11 − 8 | 11 − 6 | 11 − 4 | 11 − 9 | 11 − 7 |

## 2

**A** 15 { 9 ______ ; ☐ ______

**B** 13 { 9 ______ ; ☐ ______

**C** 17 { 9 ______ ; ☐ ______

**D** 16 { 9 ______ ; ☐ ______

## 3

| | | | | | | | |
|---|---|---|---|---|---|---|---|
| 12 − 6 | 15 − 6 | 13 − 9 | 14 − 5 | 16 − 7 | 10 − 5 | 15 − 6 | 12 − 9 |
| 17 − 8 | 16 − 9 | 16 − 10 | 15 − 6 | 13 − 9 | 14 − 5 | 12 − 9 | 17 − 8 |

## 4

| | | | | | | | |
|---|---|---|---|---|---|---|---|
| 13 − 5 | 8 − 2 | 9 − 3 | 13 − 7 | 9 − 5 | 13 − 6 | 15 − 8 | 15 − 7 |
| 13 − 7 | 9 − 4 | 13 − 5 | 9 − 6 | 16 − 7 | 13 − 8 | 9 − 3 | 15 − 7 |

Part 4 continues on the next page.

$$\begin{array}{r} 9 \\ -7 \\ \hline \end{array} \quad \begin{array}{r} 15 \\ -\ 8 \\ \hline \end{array} \quad \begin{array}{r} 8 \\ -3 \\ \hline \end{array} \quad \begin{array}{r} 13 \\ -\ 6 \\ \hline \end{array} \quad \begin{array}{r} 9 \\ -2 \\ \hline \end{array} \quad \begin{array}{r} 8 \\ -5 \\ \hline \end{array} \quad \begin{array}{r} 15 \\ -\ 7 \\ \hline \end{array} \quad \begin{array}{r} 14 \\ -\ 5 \\ \hline \end{array}$$

$$\begin{array}{r} 9 \\ -4 \\ \hline \end{array} \quad \begin{array}{r} 13 \\ -\ 7 \\ \hline \end{array} \quad \begin{array}{r} 13 \\ -\ 6 \\ \hline \end{array} \quad \begin{array}{r} 15 \\ -\ 8 \\ \hline \end{array} \quad \begin{array}{r} 13 \\ -\ 8 \\ \hline \end{array} \quad \begin{array}{r} 10 \\ -\ 6 \\ \hline \end{array} \quad \begin{array}{r} 9 \\ -4 \\ \hline \end{array} \quad \begin{array}{r} 6 \\ -2 \\ \hline \end{array}$$

**A** Jana and Allison like to play marbles. Jana has 137 marbles. Allison has 170 marbles. How many more marbles does Allison have than Jana?

**B** My mother's new car weighs 1340 kilograms. My mother's old car weighed 1700 kilograms. How many kilograms do the cars weigh in all?

**C** Mr. Mills is 45 years old. Mr. Young is 60 years old. How many years older is Mr. Young?

**D** Rafael and Matt worked as ticket sellers at a circus. Rafael sold 94 tickets and Matt sold 120 tickets. How many tickets did the boys sell?

**E** An apple tree is 594 centimeters tall. An oak tree is 730 centimeters tall. How many centimeters taller than the apple tree is the oak tree?

**A** The store sold 185 red pencils. It sold 480 black pencils. How many pencils did the store sell?

[ ] ____________________

Part 6 continues on the next page.

**B** 154 cooks entered a cooking contest. 89 cooks won prizes. How many cooks did not win prizes?

[ ] ________________

**C** In a large market, shoppers were asked to taste a new kind of cake. 425 shoppers liked the cake. 128 shoppers did not like the cake. How many shoppers tasted the cake?

[ ] ________________

**D** When a zoo got Emma the elephant, she weighed 2589 kilograms. Now Emma weighs 3420 kilograms. How many kilograms has Emma gained since she has been in the zoo?

[ ] ________________

**E** There were 1365 boys at the school bicycle races. There were 2100 children at these races. How many of them were girls?

[ ] ________________

**F** A farmer sent 548 potatoes to a market in the city. He also sent 379 onions. How many vegetables did the farmer send?

[ ] ________________

**G** Ann likes to read mystery stories. In June she read 28 mystery stories. In July she read 17 mystery stories. How many mystery stories did she read?

[ ] ________________

Part 6 continues on the next page.

**H** Last year our family took 1444 photographs. 180 of them were black and white. How many of the photographs were not black and white?

**I** There were 518 people at a dance. Some more people came to the dance. Then there were 590 people at the dance. How many more people came to the dance?

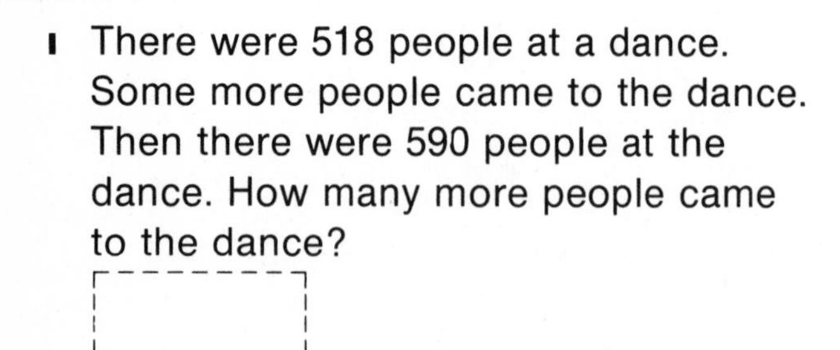

**J** 42 ducks landed on the pond on Sunday. On Tuesday 30 ducks landed on the pond. How many ducks landed on the pond?

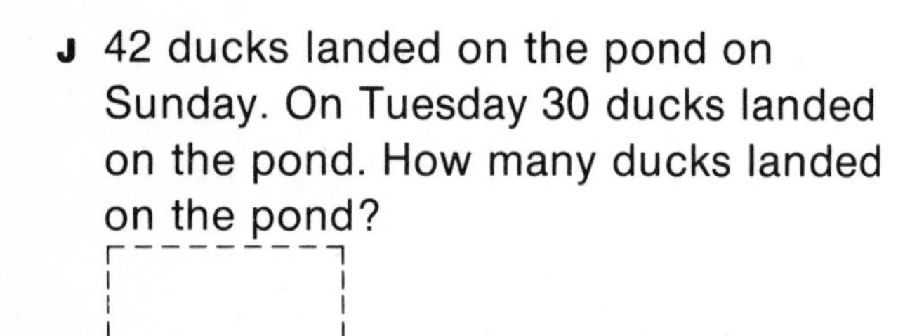

**K** 520 people attended a concert. 34 people went home early. How many people stayed at the concert?

## 7

| | | | | |
|---|---|---|---|---|
| **A** $\begin{array}{r} 6003 \\ -\ \ 80 \\ \hline \end{array}$ | **B** $\begin{array}{r} 4200 \\ -\ 905 \\ \hline \end{array}$ | **C** $\begin{array}{r} 1800 \\ -\ 286 \\ \hline \end{array}$ | **D** $\begin{array}{r} 1000 \\ -\ 888 \\ \hline \end{array}$ | **E** $\begin{array}{r} 5006 \\ -\ 206 \\ \hline \end{array}$ |
| **F** $\begin{array}{r} 3102 \\ -1996 \\ \hline \end{array}$ | **G** $\begin{array}{r} 1000 \\ -\ 200 \\ \hline \end{array}$ | **H** $\begin{array}{r} 4105 \\ -3999 \\ \hline \end{array}$ | **I** $\begin{array}{r} 6000 \\ -4990 \\ \hline \end{array}$ | **J** $\begin{array}{r} 5000 \\ -3500 \\ \hline \end{array}$ |

# Lesson 57

| Facts | + | Problems | + | Bonus | = | TOTAL |
|---|---|---|---|---|---|---|
| | | | | | | |

**1**

| | | | | | | | |
|---|---|---|---|---|---|---|---|
| 15 − 8 | 15 − 6 | 15 − 9 | 15 − 7 | 15 − 10 | 15 − 6 | 15 − 9 | 15 − 8 |
| 15 − 10 | 15 − 7 | 15 − 9 | 15 − 8 | 15 − 6 | 15 − 7 | 15 − 8 | 15 − 6 |

**2**

**A** 13 { 9 ______ ; ☐ ______

**B** 17 { 9 ______ ; ☐ ______

**C** 14 { 9 ______ ; ☐ ______

**D** 11 { 9 ______ ; ☐ ______

**3**

| | | | | | | | |
|---|---|---|---|---|---|---|---|
| 14 − 5 | 16 − 8 | 15 − 9 | 13 − 4 | 17 − 9 | 14 − 7 | 17 − 8 | 15 − 8 |
| 14 − 5 | 11 − 9 | 13 − 9 | 17 − 8 | 16 − 8 | 12 − 9 | 15 − 7 | 17 − 8 |

**4**

| | | | | | | | |
|---|---|---|---|---|---|---|---|
| 11 − 7 | 13 − 9 | 15 − 6 | 11 − 8 | 13 − 5 | 17 − 8 | 13 − 7 | 11 − 6 |
| 13 − 5 | 11 − 9 | 15 − 7 | 13 − 4 | 11 − 6 | 13 − 6 | 13 − 8 | 11 − 7 |

Part 4 continues on the next page.

| | | | | | | | |
|---|---|---|---|---|---|---|---|
| 9 − 3 | 11 − 8 | 9 − 7 | 11 − 4 | 8 − 3 | 11 − 5 | 9 − 5 | 11 − 3 |
| 8 − 5 | 16 − 9 | 11 − 2 | 9 − 6 | 7 − 3 | 9 − 4 | 13 − 6 | 9 − 4 |

## 5

**A** We saw two horses pulling a wagon. Their owner said that one weighed 874 kilograms. The other weighed 920 kilograms. How many kilograms did the horses weigh in all?

[ ] ________________

**B** Our city has 46 restaurants and 90 shops. How many more shops does the city have than restaurants?

[ ] ________________

**C** The officers who keep track of animals in a national park put tags on 1482 deer and 1800 elk. How many tags were put on in all?

[ ] ________________

**D** A young elephant weighs 1350 kilograms. Its mother weighs 2500 kilograms. How many more kilograms does the mother elephant weigh?

[ ] ________________

**E** Mr. Welbes carved 62 wooden ducks. Mr. Kiyo carved 39 wooden ducks. How many more ducks did Mr. Welbes carve than Mr. Kiyo?

[ ] ________________

## 6

**A** Mr. Gallo planted 147 fruit trees on his ranch. 28 fruit trees did not grow. How many fruit trees did grow?

[ ] ____________

**B** Last week a baker sold 1350 blueberry muffins. This week the baker sold 1500 cherry muffins. How many muffins did the baker sell?

[ ] ____________

**C** Mr. Delgado teaches piano. He had 167 students. Then he got some more students. Now he has 210 students. How many more students did he get?

[ ] ____________

**D** 190 children came to school today. 175 of the children ate a hot lunch. How many children did not eat a hot lunch?

[ ] ____________

**E** Tree cutters were cutting down pine trees and maple trees. They loaded 185 pine logs and 190 maple logs onto railway cars. How many logs in all were loaded?

[ ] ____________

**F** At the automobile races there were 1490 men. 2100 persons went to the races. How many were women?

[ ] ____________

Part 6 continues on the next page.

**G** Maria did 97 subtraction problems. Then she did some more problems. Now she has done 102 problems. How many more problems did she do?

[ ] ____________

**H** Last summer Mike and Holly counted 158 crows and 190 sparrows. How many birds did they count?

[ ] ____________

**I** A men's shop had 1500 shirts on sale. The shop sold 1389 of the shirts. How many shirts does the shop have now?

[ ] ____________

**J** Our family moved to Bellwood. We had 143 large boxes and 180 small boxes. How many boxes did we have?

[ ] ____________

## 7

**A** $5000 - 386$

**B** $3400 - 280$

**C** $7400 - 1648$

**D** $5100 - 243$

**E** $6000 - 5940$

**F** $3100 - 264$

**G** $5000 - 4138$

**H** $6300 - 680$

**I** $1000 - 349$

**J** $3010 - 999$

# Lesson 58

| Test | + | Facts | + | Problems | + | Bonus | = | TOTAL |
|---|---|---|---|---|---|---|---|---|
| | | | | | | | | |

## 1

| 15 − 7 | 15 − 6 | 15 − 8 | 15 − 10 | 15 − 9 | 15 − 6 | 15 − 10 | 15 − 8 |
|---|---|---|---|---|---|---|---|
| 15 − 9 | 15 − 7 | 15 − 6 | 15 − 9 | 15 − 8 | 15 − 7 | 15 − 9 | 15 − 6 |

## 2

**A** 14 { 9 ______ ; [ ] ______

**B** 13 { 9 ______ ; [ ] ______

**C** 17 { 9 ______ ; [ ] ______

**D** 16 { 9 ______ ; [ ] ______

## 3

| 17 − 8 | 14 − 5 | 14 − 7 | 16 − 8 | 17 − 8 | 12 − 9 | 13 − 4 | 16 − 9 |
|---|---|---|---|---|---|---|---|
| 16 − 7 | 14 − 9 | 17 − 8 | 12 − 6 | 14 − 7 | 16 − 7 | 16 − 9 | 14 − 5 |

## 4

| 11 − 6 | 13 − 8 | 16 − 7 | 11 − 8 | 13 − 4 | 13 − 5 | 11 − 7 | 14 − 5 |
|---|---|---|---|---|---|---|---|
| 11 − 3 | 13 − 6 | 11 − 5 | 9 − 2 | 11 − 4 | 9 − 4 | 14 − 8 | 13 − 7 |

Part 4 continues on the next page.

$$\begin{array}{r} 6 \\ -\ 2 \\ \hline \end{array} \quad \begin{array}{r} 11 \\ -\ 7 \\ \hline \end{array} \quad \begin{array}{r} 9 \\ -\ 5 \\ \hline \end{array} \quad \begin{array}{r} 13 \\ -\ 6 \\ \hline \end{array} \quad \begin{array}{r} 11 \\ -\ 6 \\ \hline \end{array} \quad \begin{array}{r} 8 \\ -\ 3 \\ \hline \end{array} \quad \begin{array}{r} 9 \\ -\ 6 \\ \hline \end{array} \quad \begin{array}{r} 11 \\ -\ 8 \\ \hline \end{array}$$

$$\begin{array}{r} 7 \\ -\ 3 \\ \hline \end{array} \quad \begin{array}{r} 11 \\ -\ 2 \\ \hline \end{array} \quad \begin{array}{r} 9 \\ -\ 3 \\ \hline \end{array} \quad \begin{array}{r} 8 \\ -\ 2 \\ \hline \end{array} \quad \begin{array}{r} 11 \\ -\ 9 \\ \hline \end{array} \quad \begin{array}{r} 8 \\ -\ 5 \\ \hline \end{array} \quad \begin{array}{r} 14 \\ -\ 6 \\ \hline \end{array} \quad \begin{array}{r} 12 \\ -\ 7 \\ \hline \end{array}$$

## 5

**A** Mrs. Oliver is 42 years old. Mr. Oliver is 60 years old. How many years younger is Mrs. Oliver?

[ ] ________________

**B** There were two hippopotamuses. The smaller one weighed 1342 kilograms. The larger hippopotamus weighed 1520 kilograms. How many kilograms lighter is the smaller hippopotamus?

[ ] ________________

**C** The Ace Trucking Company has two trucks. The first truck was driven 1480 kilometers. The second truck was driven 1520 kilometers. How many kilometers in all were the trucks driven?

[ ] ________________

**D** Two bridges are in nearby cities. The first bridge is 426 meters long. The second bridge is 510 meters long. How much longer is the second bridge than the first bridge?

[ ] ________________

**E** Elisa picked 142 carrots from her garden. Her sister Melissa picked 180 carrots from her garden. How many carrots did both sisters pick from the gardens?

[ ] ________________

## 6

**A** Westchester had 438 buses. Then the city bought some more buses. Now there are 466 buses. How many more buses did the city buy?

[ ] ____________

**B** There are 18 red wires and 24 black wires in the TV set. How many wires are there?

[ ] ____________

**C** 1420 girls went to a parade. 1700 children went to the parade. How many boys went to the parade?

[ ] ____________

**D** Mrs. Garcia ordered 164 bottles for her shop. 98 of them were broken. How many of the bottles were not broken?

[ ] ____________

**E** A factory built 2145 cars that had radios and 3190 cars that did not have radios. How many cars in all did the factory build?

[ ] ____________

**F** There were 95 people in a restaurant. Later, more people came to the restaurant. Then there were 120 people in the restaurant. How many people came later?

[ ] ____________

Part 6 continues on the next page.

**G** 1500 runners were asked to try a new kind of running shoe. 864 runners did not like them. How many runners liked the new kind of shoe?

[ ] ______________

**H** In an exercise class 124 girls are jumping rope. 150 girls are swimming. How many girls are in the class?

[ ] ______________

**I** 148 people were watching Eric shoot his bow and arrow. 52 more people joined them. How many people were watching Eric then?

[ ] ______________

**J** On Monday 1405 boxes of blueberries were for sale at a market. By Friday 1289 boxes of blueberries had been sold. How many boxes were left for the market to sell?

[ ] ______________

## 7

**A** $4600 - 2780$

**B** $1000 - 386$

**C** $7043 - 5239$

**D** $4100 - 3683$

**E** $6000 - 320$

**F** $5310 - 4298$

**G** $4000 - 3369$

**H** $2400 - 908$

**I** $7100 - 6936$

**J** $4100 - 62$

# Lesson 59

| Facts | + | Problems | + | Bonus | = | TOTAL |
|---|---|---|---|---|---|---|
| | | | | | | |

## 1

| 13 | 13 | 16 | 12 | 12 | 14 | 14 | 14 |
|---|---|---|---|---|---|---|---|
| − 9 | − 6 | − 7 | − 6 | − 9 | − 7 | − 5 | − 9 |

| 18 | 15 | 15 | 13 | 13 | 17 | 17 | 15 |
|---|---|---|---|---|---|---|---|
| − 9 | − 6 | − 9 | − 4 | − 9 | − 8 | − 9 | − 6 |

## 2

**A**

8 { 2 ________ ; ☐ ________

**B**

6 { 2 ________ ; ☐ ________

## 3

| 6 | 11 | 13 | 11 | 8 | 13 | 15 | 11 |
|---|---|---|---|---|---|---|---|
| − 4 | − 7 | − 5 | − 8 | − 6 | − 4 | − 6 | − 6 |

| 13 | 11 | 13 | 8 | 11 | 11 | 13 | 6 |
|---|---|---|---|---|---|---|---|
| − 6 | − 4 | − 5 | − 6 | − 6 | − 3 | − 8 | − 4 |

| 15 | 11 | 13 | 11 | 8 | 11 | 11 | 13 |
|---|---|---|---|---|---|---|---|
| − 8 | − 2 | − 6 | − 5 | − 6 | − 2 | − 7 | − 5 |

| 9 | 13 | 6 | 15 | 15 | 11 | 15 | 15 |
|---|---|---|---|---|---|---|---|
| − 2 | − 6 | − 4 | − 7 | − 9 | − 8 | − 8 | − 6 |

## 4

**A** Two camels were carrying food across the desert. A brown camel carried 126 kilograms of food. A gray camel carried 140 kilograms of food. How many kilograms lighter was the food that the brown camel carried?

☐ ________

Part 4 continues on the next page.

**B** Steve and Kelly collect gold coins. Steve has 39 coins. Kelly has 46 coins. How many fewer coins does Steve have?

[ ] ________

**C** Bart and Roy raise chickens. Bart has 47 chickens and Roy has 92 chickens. How many chickens in all do they have?

[ ] ________

**D** A card shop sold 1400 get-well cards and 1900 birthday cards last month. How many more birthday cards did the shop sell than get-well cards?

[ ] ________

**E** Ian saw many boats on the lake as he rode his bicycle. He saw 85 sailboats and 29 motorboats. How many boats in all did Ian see?

[ ] ________

**F** Cathy has 147 joke books. Joe has 210 joke books. How many more joke books does Joe have than Cathy?

[ ] ________

## 5

**A** The Happy Company has 90 pieces of pipe. 72 pieces of pipe are straight. How many pieces of pipe are not straight?

[ ] ________

**B** We went walking in the woods. We saw 143 field mice and 180 rabbits. How many animals did we see?

[ ] ________

Part 5 continues on the next page.

**C** Mr. Banaka is a champion swimmer. He earned 1842 points for his swimming skill during the first hour of a contest. Then he got 88 more points for diving. How many points in all did Mr. Banaka get?

[ ] ________________

**D** A soccer team has won 83 games in the last four years. The team has played 90 games. How many games has the team lost?

[ ] ________________

**E** A hotel bought 1444 light bulbs. 98 of them were broken. How many light bulbs were not broken?

[ ] ________________

**F** 136 people were riding horses. 96 people got off of their horses. How many people are still on their horses?

[ ] ________________

**G** A book shop sold 1826 books in December. The store sold 1900 books in January. How many books did the shop sell?

[ ] ________________

**H** 45 cows are in the barn. 90 cows are not in the barn. How many cows are there in all?

[ ] ________________

Part 5 continues on the next page.

**I** When a cooking school opened, 148 people signed up to take classes. Other people came later to sign up for classes. Then there were 190 students in the cooking school. How many people came later?

[ ] ____________________

**6**

**A** $3100 - 86$

**B** $7000 - 230$

**C** $4000 - 8$

**D** $1000 - 20$

**E** $6200 - 4936$

**F** $1000 - 923$

**G** $5100 - 486$

**H** $2000 - 180$

**I** $7200 - 943$

**J** $6010 - 4798$

# Lesson 60

Test + Facts + Problems + Bonus = TOTAL

## 1

| | | | | | | | |
|---|---|---|---|---|---|---|---|
| 13 − 5 | 13 − 7 | 13 − 4 | 13 − 6 | 13 − 8 | 13 − 9 | 13 − 5 | 13 − 4 |
| 13 − 6 | 13 − 8 | 13 − 9 | 13 − 5 | 13 − 6 | 13 − 7 | 13 − 4 | 13 − 8 |

## 2

**A** 6 { 2 ______ ; ☐ ______

**B** 8 { 2 ______ ; ☐ ______

## 3

| | | | | | | | |
|---|---|---|---|---|---|---|---|
| 17 − 8 | 8 − 6 | 11 − 6 | 15 − 6 | 11 − 8 | 14 − 5 | 6 − 4 | 11 − 7 |
| 15 − 6 | 11 − 5 | 15 − 7 | 11 − 2 | 13 − 4 | 15 − 8 | 11 − 6 | 15 − 9 |
| 17 − 8 | 11 − 3 | 9 − 2 | 15 − 7 | 11 − 8 | 15 − 6 | 11 − 4 | 15 − 8 |
| 14 − 5 | 11 − 2 | 10 − 4 | 11 − 7 | 16 − 7 | 12 − 3 | 8 − 3 | 12 − 8 |

## 4

**A** Our house is 42 years old. The house next door is 60 years old. How many years newer is our house?

☐ ______

Part 4 continues on the next page.

**B** There are 46 white turkeys and 90 gray turkeys on a farm. How many more gray turkeys are there than white ones?

**C** A carnival sold 94 balloons on the first day. On the second day the carnival sold 120 balloons. How many balloons did the carnival sell in all?

**D** There were two huge statues in a park. One weighed 1340 kilograms. The other weighed 1700 kilograms. How many kilograms did the statues weigh in all?

**E** A fruit grower sent 1400 boxes of lemons and 1900 boxes of oranges to a market. How many more boxes of oranges were sent than lemons?

## 5

**A** Mr. Jenssen bought an old chair that was 37 years old. Mr. Merino bought a sofa that was 50 years old. How many years older was the sofa?

**B** In a parade we saw 150 police officers riding horses and 150 cowboys riding horses. How many people in all were riding horses?

Part 5 continues on the next page.

**C** Our city has 1960 green taxis and 3045 red taxis. How many taxis does our city have?

**D** Louis and Girard were writing a book. In May they wrote 175 pages. In June they wrote 190 pages. How many pages did they write in May and June?

**E** A huge oak desk weighs 124 kilograms. A brass lamp on the desk weighs 48 kilograms. How much do the desk and lamp weigh altogether?

**F** Our basketball team won 47 games. The team played 52 games in all. How many games did the team lose?

**G** 42 trains went through the tunnel on Sunday. 38 trains went through the tunnel on Monday. How many trains went through the tunnel?

**H** 154 passengers were on a plane that was going to China. In Los Angeles 99 people got off. How many people did not get off?

**I** A men's shop sold 85 plain neckties and 139 striped neckties. How many neckties did the shop sell?

Part 5 continues on the next page.

**J** Mr. Johnson has a newsstand. He sold 138 newspapers in the morning. Then he sold some more newspapers in the afternoon. Now he has sold 150 newspapers. How many newspapers did he sell in the afternoon?

[ ] ____________________

## 6

**A** $\begin{array}{r} 4000 \\ -\ 1236 \\ \hline \end{array}$

**B** $\begin{array}{r} 8100 \\ -\ 26 \\ \hline \end{array}$

**C** $\begin{array}{r} 4030 \\ -\ 594 \\ \hline \end{array}$

**D** $\begin{array}{r} 6000 \\ -\ 4123 \\ \hline \end{array}$

**E** $\begin{array}{r} 5200 \\ -\ 186 \\ \hline \end{array}$

**F** $\begin{array}{r} 3100 \\ -\ 84 \\ \hline \end{array}$

**G** $\begin{array}{r} 1000 \\ -\ 90 \\ \hline \end{array}$

**H** $\begin{array}{r} 4000 \\ -\ 1555 \\ \hline \end{array}$

**I** $\begin{array}{r} 3600 \\ -\ 280 \\ \hline \end{array}$

**J** $\begin{array}{r} 4126 \\ -\ 3996 \\ \hline \end{array}$

# Lesson 61

Facts + Problems + Bonus = TOTAL

## 1

$13 - 6$ $13 - 4$ $13 - 9$ $13 - 5$ $13 - 6$ $13 - 8$ $13 - 7$ $13 - 9$

$13 - 6$ $13 - 4$ $13 - 7$ $13 - 5$ $13 - 8$ $13 - 9$ $13 - 6$ $13 - 4$

## 2

$6 - 4$ $11 - 7$ $17 - 8$ $11 - 6$ $15 - 7$ $14 - 5$ $8 - 6$ $11 - 8$

$11 - 4$ $16 - 7$ $11 - 9$ $15 - 8$ $14 - 6$ $13 - 4$ $8 - 6$ $11 - 5$

$11 - 3$ $14 - 5$ $6 - 4$ $11 - 8$ $15 - 6$ $11 - 7$ $15 - 7$ $11 - 4$

$17 - 8$ $15 - 8$ $9 - 4$ $8 - 3$ $13 - 4$ $8 - 5$ $12 - 5$ $15 - 6$

## 3

**A** There are 170 birds in a zoo. There are 200 fish in the zoo. How many fewer birds than fish are in the zoo?

[ ] ________________

**B** We saw lots of squirrels and rabbits. We counted 143 squirrels and 180 rabbits. How many animals did we see?

[ ] ________________

Part 3 continues on the next page.

**C** The cook let us choose cake or ice cream for dessert. 145 kids ate cake and 200 kids ate ice cream. How many more kids ate ice cream than cake?

[ ] ________________

**D** Iris taught 36 children how to jump rope. Christina taught 40 children how to jump rope. How many children learned how to jump rope?

[ ] ________________

**E** Chris has 47 rocks. Judd has 82 rocks. How many more rocks does Judd have than Chris?

[ ] ________________

**F** There are 90 bottles of soda pop. 72 of the bottles are full. How many bottles are not full?

[ ] ________________

**G** 1350 railway cars are full. 1500 railway cars are empty. How many railway cars are there in all?

[ ] ________________

**H** The Fireside Bookshop sold 1826 books in December. The shop sold 1900 books in January. How many books did the shop sell?

[ ] ________________

**I** A hockey team lost 42 games. It played 70 games in all. How many games did the hockey team win?

[ ] ________________

Part 3 continues on the next page.

**J** Fran ran 37 kilometers this week. Carol ran 40 kilometers this week. How many kilometers did Fran and Carol run this week?

[ ] ____________________

**K** Louise made 49 small cakes with strawberry frosting. Then she made some cakes with chocolate frosting. She baked 72 cakes in all. How many had chocolate frosting?

[ ] ____________________

**L** Ann read 167 pages of a book. On her day off she read some more. Now Ann has read 210 pages. How many pages did Ann read on her day off?

[ ] ____________________

## 4

| A | B | C | D | E |
|---|---|---|---|---|
| 3010 − 826 | 5200 + 380 | 6000 − 84 | 4106 − 3289 | 5006 − 95 |

| F | G | H | I | J |
|---|---|---|---|---|
| 8105 − 7999 | 5010 − 4364 | 8246 − 7580 | 9000 − 4280 | 6400 + 287 |

| K | L | M | N | O |
|---|---|---|---|---|
| 3200 − 1800 | 5130 − 4830 | 9042 + 186 | 7036 − 189 | 1000 − 285 |

# Lesson 62

Facts + Problems + Bonus = TOTAL

**1**

A

$$17 \begin{cases} 9 \\ \square \end{cases}$$

B

$$14 \begin{cases} 9 \\ \square \end{cases}$$

C

$$15 \begin{cases} 9 \\ \square \end{cases}$$

D

$$16 \begin{cases} 9 \\ \square \end{cases}$$

**2**

| 15 − 9 | 17 − 8 | 12 − 6 | 10 − 5 | 16 − 7 | 14 − 7 | 14 − 9 | 15 − 6 |
|---|---|---|---|---|---|---|---|
| 15 − 7 | 14 − 5 | 14 − 6 | 14 − 7 | 17 − 8 | 17 − 9 | 16 − 7 | 14 − 5 |

**3**

| 14 − 6 | 17 − 8 | 9 − 4 | 13 − 8 | 8 − 6 | 15 − 7 | 6 − 4 | 11 − 4 |
|---|---|---|---|---|---|---|---|
| 15 − 7 | 13 − 7 | 15 − 6 | 8 − 6 | 11 − 8 | 6 − 4 | 14 − 5 | 11 − 3 |
| 15 − 8 | 9 − 5 | 6 − 4 | 13 − 6 | 15 − 8 | 12 − 3 | 13 − 7 | 11 − 2 |
| 13 − 4 | 14 − 9 | 13 − 5 | 8 − 5 | 13 − 8 | 17 − 8 | 15 − 7 | 13 − 6 |

## 4

**A** 64 people walked around the ladder. 140 people walked under the ladder. How many more people walked under the ladder than around the ladder?

[ ] ____________

**B** In one week 148 new cars and 473 old cars were washed at the Happy Car Wash. How many cars in all were washed?

[ ] ____________

**C** At a camp 3802 oranges were eaten in one week. 1648 apples were also eaten. How many more oranges than apples were eaten?

[ ] ____________

**D** Andy put 2509 boxes on trucks. Randy put 3108 boxes on trucks. How many fewer boxes did Andy put on trucks than Randy?

[ ] ____________

**E** Workers chopped down 185 oak trees and 190 pine trees. How many trees were chopped down?

[ ] ____________

**F** Last spring a shop had a sale on kites. The shop sold 158 red kites and 190 yellow kites. How many kites did the shop sell?

[ ] ____________

Part 4 continues on the next page.

**G** Lauren planted 97 flowers. Then she planted some more flowers. Now 102 flowers are planted. How many more flowers did she plant?

[ ] ________________

**H** A newspaper stand received 1444 magazines. 98 of the magazines had torn pages. How many magazines did not have torn pages?

[ ] ________________

**I** 148 people were ice-skating. Later, more people came to skate. Now there are 190 skaters. How many people came late to skate?

[ ] ________________

**J** Phil danced for 148 minutes. Then he danced for another 52 minutes. How many minutes did he dance in all?

[ ] ________________

**K** Workers cleaned 18 airplanes. They cleaned 24 boats. How many things did the workers clean?

[ ] ________________

**L** An office had 1405 packages of paper. The office used up 1289 packages. How many packages of paper were left?

[ ] ________________

**A** $\begin{array}{r} 5000 \\ -\ 4836 \\ \hline \end{array}$

**B** $\begin{array}{r} 6200 \\ +\ \ 886 \\ \hline \end{array}$

**C** $\begin{array}{r} 5100 \\ -\ \ 284 \\ \hline \end{array}$

**D** $\begin{array}{r} 3200 \\ -\ 1964 \\ \hline \end{array}$

**E** $\begin{array}{r} 1000 \\ -\ \ \ 20 \\ \hline \end{array}$

**F** $\begin{array}{r} 3400 \\ -\ 1950 \\ \hline \end{array}$

**G** $\begin{array}{r} 5010 \\ -\ 4286 \\ \hline \end{array}$

**H** $\begin{array}{r} 3825 \\ +\ \ 182 \\ \hline \end{array}$

**I** $\begin{array}{r} 9306 \\ -\ \ 829 \\ \hline \end{array}$

**J** $\begin{array}{r} 3000 \\ -\ 1280 \\ \hline \end{array}$

| Test | + | Facts | + | Problems | + | Bonus | = | TOTAL |
|---|---|---|---|---|---|---|---|---|
| | | | | | | | | |

**1**

| | | | | | | | |
|---|---|---|---|---|---|---|---|
| $14 - 5$ | $14 - 9$ | $14 - 7$ | $17 - 8$ | $17 - 9$ | $16 - 8$ | $15 - 6$ | $13 - 4$ |
| $18 - 9$ | $14 - 7$ | $16 - 7$ | $14 - 5$ | $12 - 6$ | $17 - 8$ | $17 - 9$ | $16 - 7$ |

**2**

| | | | | | | | |
|---|---|---|---|---|---|---|---|
| $14 - 5$ | $11 - 7$ | $13 - 6$ | $11 - 9$ | $16 - 7$ | $15 - 7$ | $11 - 8$ | $11 - 6$ |
| $13 - 7$ | $13 - 4$ | $15 - 8$ | $11 - 4$ | $13 - 5$ | $12 - 3$ | $13 - 4$ | $11 - 3$ |
| $13 - 8$ | $17 - 8$ | $11 - 5$ | $13 - 7$ | $15 - 6$ | $11 - 7$ | $13 - 5$ | $11 - 4$ |
| $8 - 3$ | $11 - 2$ | $8 - 6$ | $11 - 8$ | $6 - 4$ | $14 - 5$ | $8 - 3$ | $9 - 4$ |

**3**

**A** Martin and Frank were selling balloons. Martin sold 93 balloons. Frank sold 130 balloons. How many fewer balloons did Martin sell than Frank?

[ ] ____________________

Part 3 continues on the next page.

**B** 425 people voted for Sam. 128 people did not vote for Sam. How many people voted in all?

[ ] __________________

**C** 195 people came to the meeting early. 24 people came late. How many people in all came to the meeting?

[ ] __________________

**D** Rob has 36 rabbits. Tony has 52 rabbits. How many more rabbits does Tony have than Rob?

[ ] __________________

**E** There were 154 fish. 89 of the fish were swordfish. How many of the fish were not swordfish?

[ ] __________________

**F** In a building 85 windows were closed. 17 windows were open. How many windows did the building have?

[ ] __________________

**G** A sign outside an office building was 48 meters tall. It stood next to a pole that was 60 meters tall. How many meters taller was the pole?

[ ] __________________

**H** In our city there are 548 brick buildings and 379 wooden buildings. How many buildings are in our city?

[ ] __________________

Part 3 continues on the next page.

**I** 143 planes landed late. 583 planes did not land late. How many planes landed in all?

☐ ________________

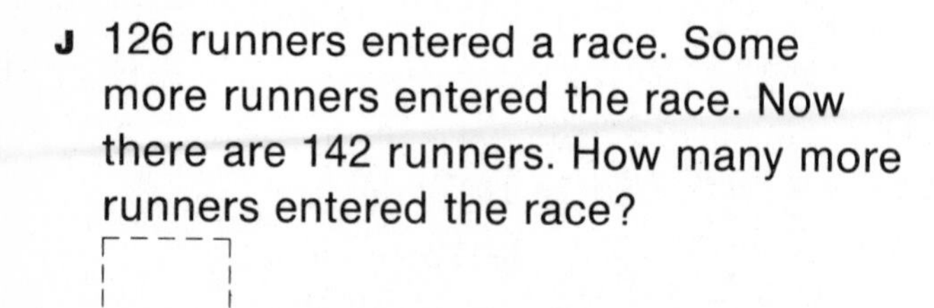

**J** 126 runners entered a race. Some more runners entered the race. Now there are 142 runners. How many more runners entered the race?

☐ ________________

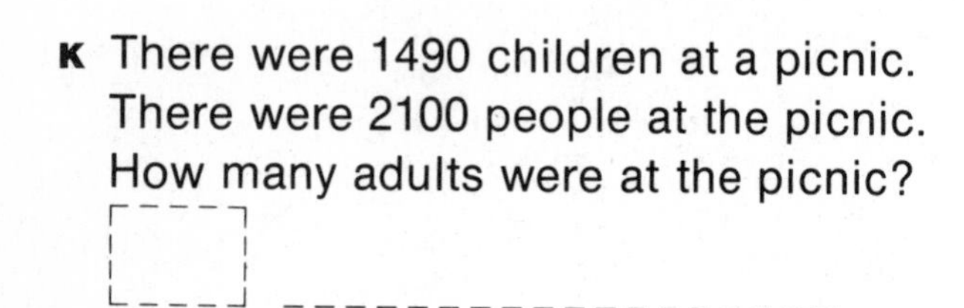

**K** There were 1490 children at a picnic. There were 2100 people at the picnic. How many adults were at the picnic?

☐ ________________

**L** 1500 people went to a flower show. 864 people did not buy flowers and plants. How many people did buy flowers and plants?

☐ ________________

## 4

| A | B | C | D | E |
|---|---|---|---|---|
| $3200 - 864$ | $6400 - 4980$ | $6000 - 37$ | $1280 + 880$ | $7100 - 3489$ |

| F | G | H | I | J |
|---|---|---|---|---|
| $6010 - 999$ | $1000 - 640$ | $1000 - 39$ | $6100 - 436$ | $9500 - 3820$ |

# Lesson 64

| Test | + | Facts | + | Bonus | = | TOTAL |
|---|---|---|---|---|---|---|
| | | | | | | |

## 1

| | | | | | | | |
|---|---|---|---|---|---|---|---|
| 15 − 6 | 15 − 9 | 14 − 7 | 14 − 5 | 14 − 9 | 16 − 8 | 17 − 8 | 16 − 8 |
| 16 − 7 | 16 − 9 | 16 − 8 | 13 − 4 | 13 − 9 | 15 − 6 | 15 − 9 | 16 − 8 |

## 2

| | | | | | | | |
|---|---|---|---|---|---|---|---|
| 11 − 5 | 14 − 5 | 6 − 4 | 15 − 7 | 11 − 8 | 13 − 5 | 17 − 8 | 11 − 7 |
| 13 − 8 | 15 − 7 | 13 − 6 | 11 − 4 | 15 − 8 | 11 − 6 | 8 − 6 | 11 − 3 |
| 13 − 7 | 10 − 3 | 11 − 2 | 13 − 5 | 16 − 7 | 15 − 8 | 6 − 4 | 13 − 8 |
| 14 − 9 | 13 − 6 | 9 − 4 | 8 − 6 | 13 − 7 | 12 − 4 | 13 − 5 | 12 − 8 |

## 3

**A** Art is 48 years old. Troy is 70 years old. How many years younger is Art?

[ ] ____________________

**B** A spaceship went around the earth 34 times the first week. During the second week the spaceship went around the earth 30 times. How many times did it go around the earth?

[ ] ____________________

Part 3 continues on the next page.

**C** A cook used 185 brown eggs and 480 white eggs. How many eggs did the cook use?

[ ] ________________

**D** A city has 148 old buses and 172 new buses. How many more new buses than old buses are there?

[ ] ________________

**E** Carol is 146 centimeters tall. Lupe is 164 centimeters tall. How many centimeters taller is Lupe?

[ ] ________________

**F** A new shop gave away 86 red pens and 90 blue pens. How many pens did the shop give away?

[ ] ________________

**G** Nancy has a dog that used to weigh 108 kilograms. The dog has gained 24 kilograms. How much does it weigh now?

[ ] ________________

**H** A small car weighs 1248 kilograms. A racing car weighs 1600 kilograms. How much heavier is the racing car?

[ ] ________________

**I** We had 164 nuts. 98 of the nuts had shells. How many of the nuts did not have shells?

[ ] ________________

Part 3 continues on the next page.

**J** Martha Rainwater counted 1463 ducks on the lake and 3652 ducks in the air. How many ducks did she count?

[ ] ____________________

**K** 184 girls were playing checkers. 176 boys were playing checkers. How many children were playing checkers?

[ ] ____________________

**L** At the beginning of the week Kurt had driven his car 945 kilometers. At the end of the week Kurt had driven 1200 kilometers. How many kilometers did Kurt drive his car during the week?

[ ] ____________________

# Lesson 65

| Test | + | Problems | + | Bonus | = | TOTAL |
|---|---|---|---|---|---|---|
| | | | | | | |

## 1

$$\begin{array}{cccccccc}
\begin{array}{r}6\\-4\\\hline\end{array} & \begin{array}{r}13\\-9\\\hline\end{array} & \begin{array}{r}17\\-8\\\hline\end{array} & \begin{array}{r}8\\-6\\\hline\end{array} & \begin{array}{r}10\\-4\\\hline\end{array} & \begin{array}{r}13\\-5\\\hline\end{array} & \begin{array}{r}11\\-2\\\hline\end{array} & \begin{array}{r}15\\-7\\\hline\end{array}
\end{array}$$

$$\begin{array}{cccccccc}
\begin{array}{r}11\\-8\\\hline\end{array} & \begin{array}{r}16\\-9\\\hline\end{array} & \begin{array}{r}11\\-6\\\hline\end{array} & \begin{array}{r}13\\-7\\\hline\end{array} & \begin{array}{r}10\\-6\\\hline\end{array} & \begin{array}{r}11\\-4\\\hline\end{array} & \begin{array}{r}12\\-9\\\hline\end{array} & \begin{array}{r}13\\-4\\\hline\end{array}
\end{array}$$

$$\begin{array}{cccccccc}
\begin{array}{r}8\\-2\\\hline\end{array} & \begin{array}{r}14\\-5\\\hline\end{array} & \begin{array}{r}16\\-7\\\hline\end{array} & \begin{array}{r}10\\-2\\\hline\end{array} & \begin{array}{r}13\\-6\\\hline\end{array} & \begin{array}{r}11\\-9\\\hline\end{array} & \begin{array}{r}9\\-5\\\hline\end{array} & \begin{array}{r}7\\-3\\\hline\end{array}
\end{array}$$

$$\begin{array}{cccccccc}
\begin{array}{r}13\\-8\\\hline\end{array} & \begin{array}{r}16\\-9\\\hline\end{array} & \begin{array}{r}11\\-3\\\hline\end{array} & \begin{array}{r}14\\-9\\\hline\end{array} & \begin{array}{r}8\\-5\\\hline\end{array} & \begin{array}{r}15\\-6\\\hline\end{array} & \begin{array}{r}7\\-4\\\hline\end{array} & \begin{array}{r}12\\-4\\\hline\end{array}
\end{array}$$

$$\begin{array}{cccccccc}
\begin{array}{r}9\\-3\\\hline\end{array} & \begin{array}{r}15\\-8\\\hline\end{array} & \begin{array}{r}10\\-3\\\hline\end{array} & \begin{array}{r}9\\-4\\\hline\end{array} & \begin{array}{r}9\\-7\\\hline\end{array} & \begin{array}{r}17\\-9\\\hline\end{array} & \begin{array}{r}10\\-7\\\hline\end{array} & \begin{array}{r}12\\-3\\\hline\end{array}
\end{array}$$

$$\begin{array}{cccccccc}
\begin{array}{r}14\\-6\\\hline\end{array} & \begin{array}{r}8\\-3\\\hline\end{array} & \begin{array}{r}15\\-9\\\hline\end{array} & \begin{array}{r}9\\-2\\\hline\end{array} & \begin{array}{r}11\\-5\\\hline\end{array} & \begin{array}{r}10\\-2\\\hline\end{array} & \begin{array}{r}14\\-8\\\hline\end{array} & \begin{array}{r}11\\-7\\\hline\end{array}
\end{array}$$

$$\begin{array}{cccccccc}
\begin{array}{r}12\\-4\\\hline\end{array} & \begin{array}{r}5\\-3\\\hline\end{array} & \begin{array}{r}11\\-7\\\hline\end{array} & \begin{array}{r}12\\-8\\\hline\end{array} & \begin{array}{r}12\\-5\\\hline\end{array} & \begin{array}{r}9\\-6\\\hline\end{array} & \begin{array}{r}12\\-6\\\hline\end{array} & \begin{array}{r}10\\-4\\\hline\end{array}
\end{array}$$

## 2

**A** 148 motorboats were on the river. 162 sailboats were on the river. How many more sailboats than motorboats were on the river?

[ ] ____________________

Part 2 continues on the next page.

**B** A hospital had 2365 clean sheets. It had 90 dirty sheets. How many sheets altogether did the hospital have?

[ ] ________

**C** Leslie has worked 800 hours so far this year. Holly has worked 1250 hours. How many more hours has Holly worked?

[ ] ________

**D** A jar had 438 coins in it. Some more coins were put in the jar. Now there are 466 coins in the jar. How many more coins were put in the jar?

[ ] ________

**E** 5200 men watched the play. 4800 women watched the play. How many more men watched the play than women?

[ ] ________

**F** Before eating their picnic lunch, 86 people bought milk. 14 people bought milk after lunch. How many people bought milk?

[ ] ________

**G** In our high school 124 girls are learning how to play tennis. 150 other girls are learning how to play volleyball. How many girls are learning how to play those games?

[ ] ________

Part 2 continues on the next page.

**H** Gloria is 137 centimeters tall. Ginger is 143 centimeters tall. How many centimeters taller is Ginger?

[ ] ________________

**I** A garden had 860 roses. 290 of the roses were red. How many roses were not red?

[ ] ________________

**J** Last week Mr. Kelly fixed 24 cars. This week he fixed 30 cars. How many cars has he fixed altogether?

[ ] ________________

**K** Mrs. Owens is a taxi driver. Last week she drove 436 kilometers. This week she drove 544 kilometers. How many more kilometers did she drive this week?

[ ] ________________

**L** A pet shop sold 46 puppies and 70 kittens last month. How many animals did the pet shop sell?

[ ] ________________

**M** There were 199 students in our school. Some more students came to our school. Now there are 215 students in our school. How many students came to our school?

[ ] ________________

**N** A machine has 1584 parts. 990 parts are not new. How many parts are new?

[ ] ________________

**3**

**A**
$$\begin{array}{r} 5000 \\ -\ 4820 \\ \hline \end{array}$$

**B**
$$\begin{array}{r} 6300 \\ -\ 1849 \\ \hline \end{array}$$

**C**
$$\begin{array}{r} 1000 \\ -\ \quad 87 \\ \hline \end{array}$$

**D**
$$\begin{array}{r} 5100 \\ +\ \ 826 \\ \hline \end{array}$$

**E**
$$\begin{array}{r} 9400 \\ -\ \ 826 \\ \hline \end{array}$$

**F**
$$\begin{array}{r} 7000 \\ -\ \quad 23 \\ \hline \end{array}$$

**G**
$$\begin{array}{r} 4500 \\ +\ \ 890 \\ \hline \end{array}$$

**H**
$$\begin{array}{r} 6100 \\ -\ \quad 87 \\ \hline \end{array}$$

**I**
$$\begin{array}{r} 3040 \\ -\ 2986 \\ \hline \end{array}$$

**J**
$$\begin{array}{r} 5200 \\ -\ \ 980 \\ \hline \end{array}$$

# Transition Lesson 8

BONUS

**1**

| A | B | C | D | E | F |
|---|---|---|---|---|---|
| 264 | 538 | 492 | 751 | 892 | 376 |

**2**

| A | B | C | D | E | F |
|---|---|---|---|---|---|
| 406 | 308 | 728 | 507 | 346 | 205 |

**3**

**A**

7 { 4 ______ $7 - 4 = 3$
   { 3 ______ $7 - 3 = 4$

**B**

9 { 2 ______
   { 7 ______

**C**

3 { 2 ______
   { 1 ______

**D**

5 { 3 ______
   { 2 ______

## 4

**A**

4 { 2, 2

2 + 2 = 4

4 − 2 = 2

**B**

10 { 5, 5

+

−

**C**

18 { 9, 9

+

−

**D**

6 { 3, 3

+

−

## 5

$$\begin{array}{r} 10 \\ -\ 5 \\ \hline \end{array} \quad \begin{array}{r} 6 \\ -3 \\ \hline \end{array} \quad \begin{array}{r} 6 \\ -1 \\ \hline \end{array} \quad \begin{array}{r} 18 \\ -9 \\ \hline \end{array} \quad \begin{array}{r} 6 \\ -3 \\ \hline \end{array} \quad \begin{array}{r} 8 \\ -1 \\ \hline \end{array} \quad \begin{array}{r} 7 \\ -1 \\ \hline \end{array} \quad \begin{array}{r} 18 \\ -\ 9 \\ \hline \end{array}$$

$$\begin{array}{r} 2 \\ -1 \\ \hline \end{array} \quad \begin{array}{r} 10 \\ -\ 5 \\ \hline \end{array} \quad \begin{array}{r} 6 \\ -3 \\ \hline \end{array} \quad \begin{array}{r} 8 \\ -1 \\ \hline \end{array} \quad \begin{array}{r} 18 \\ -\ 9 \\ \hline \end{array} \quad \begin{array}{r} 10 \\ -\ 5 \\ \hline \end{array} \quad \begin{array}{r} 10 \\ -\ 1 \\ \hline \end{array} \quad \begin{array}{r} 7 \\ -1 \\ \hline \end{array}$$

# Transition Lesson 25

BONUS

**1**

| A | B | C | D | E |
|---|---|---|---|---|
| 4285 | 3624 | 8153 | 9284 | 7295 |

**2**

| A | B | C | D | E | F |
|---|---|---|---|---|---|
| 4056 | 5207 | 8024 | 2902 | 4908 | 7067 |

**3**

A
$$\begin{array}{r} 435 \\ -\ 414 \\ \hline 21 \end{array}$$

B
$$\begin{array}{r} 629 \\ -\ 511 \\ \hline 18 \end{array}$$

C
$$\begin{array}{r} 5734 \\ -\ 5124 \\ \hline 610 \end{array}$$

D
$$\begin{array}{r} 7765 \\ -\ 7160 \\ \hline 605 \end{array}$$

E
$$\begin{array}{r} 867 \\ -\ 450 \\ \hline 17 \end{array}$$

**4**

A

7 { 4 $7 - 4 = 3$

3 $7 - 3 = 4$

B

9 { 2 ____________

7 ____________

C

3 { 2 ____________

1 ____________

D

5 { 3 ____________

2 ____________

## 5

A $60 - 1 =$ ______

B $70 - 1 =$ ______

C $20 - 1 =$ ______

D $40 - 1 =$ ______

E $30 - 1 =$ ______

F $50 - 1 =$ ______

G $90 - 1 =$ ______

H $80 - 1 =$ ______

## 6

A $\begin{array}{r} 3420 \\ -1801 \\ \hline \end{array}$

B $\begin{array}{r} 8486 \\ -3768 \\ \hline \end{array}$

C $\begin{array}{r} 3092 \\ -1146 \\ \hline \end{array}$

D $\begin{array}{r} 3240 \\ -1739 \\ \hline \end{array}$

## 7

A $\begin{array}{r} 752 \\ -\ \ 24 \\ \hline \end{array}$

B $\begin{array}{r} 3493 \\ -\ \ 610 \\ \hline \end{array}$

C $\begin{array}{r} 3480 \\ -\ \ \ \ 51 \\ \hline \end{array}$

D $\begin{array}{r} 70 \\ +38 \\ \hline \end{array}$

E $\begin{array}{r} 2400 \\ -\ \ \ \ 90 \\ \hline \end{array}$

F $\begin{array}{r} 3528 \\ +\ \ 528 \\ \hline \end{array}$

G $\begin{array}{r} 738 \\ +\ \ \ \ 8 \\ \hline \end{array}$

H $\begin{array}{r} 9248 \\ -1068 \\ \hline \end{array}$

I $\begin{array}{r} 7251 \\ -\ \ 601 \\ \hline \end{array}$

J $\begin{array}{r} 3846 \\ +3218 \\ \hline \end{array}$

K $\begin{array}{r} 460 \\ -355 \\ \hline \end{array}$

L $\begin{array}{r} 5284 \\ +\ \ \ \ \ \ 6 \\ \hline \end{array}$

M $\begin{array}{r} 9328 \\ -\ \ \ \ 16 \\ \hline \end{array}$

N $\begin{array}{r} 4822 \\ -\ \ 142 \\ \hline \end{array}$

O $\begin{array}{r} 8605 \\ -7150 \\ \hline \end{array}$

# Mastery Test Review—Lesson 12

**1**

| | | | | | | | |
|---|---|---|---|---|---|---|---|
| 2 − 0 | 5 − 0 | 7 − 1 | 4 − 0 | 8 − 1 | 6 − 0 | 8 − 0 | 4 − 1 |
| 3 − 0 | 8 − 1 | 6 − 0 | 4 − 0 | 10 − 1 | 3 − 1 | 6 − 0 | 2 − 0 |

# Mastery Test Review—Lesson 15

**1**

| | | | | | | | |
|---|---|---|---|---|---|---|---|
| 12 − 6 | 11 − 10 | 14 − 7 | 15 − 10 | 16 − 8 | 13 − 10 | 14 − 7 | 19 − 10 |
| 18 − 9 | 14 − 10 | 18 − 10 | 12 − 6 | 8 − 4 | 8 − 0 | 16 − 8 | 12 − 10 |
| 12 − 6 | 14 − 7 | 16 − 10 | 16 − 8 | 18 − 9 | 6 − 3 | 6 − 1 | 10 − 1 |

**2**

| | | | | | | | |
|---|---|---|---|---|---|---|---|
| 16 − 10 | 16 − 8 | 14 − 7 | 14 − 10 | 14 − 8 | 14 − 6 | 18 − 9 | 12 − 6 |
| 12 − 10 | 16 − 8 | 14 − 6 | 14 − 8 | 12 − 6 | 15 − 10 | 14 − 8 | 10 − 5 |
| 14 − 6 | 14 − 10 | 14 − 7 | 14 − 8 | 8 − 4 | 6 − 3 | 18 − 9 | 16 − 8 |

# Mastery Test Review—Lesson 22

## 1

**A**
$$\begin{array}{r} 396 \\ -248 \\ \hline \end{array}$$

**B**
$$\begin{array}{r} 607 \\ -497 \\ \hline \end{array}$$

**C**
$$\begin{array}{r} 848 \\ -670 \\ \hline \end{array}$$

**D**
$$\begin{array}{r} 942 \\ -180 \\ \hline \end{array}$$

**E**
$$\begin{array}{r} 7875 \\ -2965 \\ \hline \end{array}$$

**F**
$$\begin{array}{r} 932 \\ -826 \\ \hline \end{array}$$

**G**
$$\begin{array}{r} 3643 \\ -1840 \\ \hline \end{array}$$

**H**
$$\begin{array}{r} 5990 \\ -4189 \\ \hline \end{array}$$

**I**
$$\begin{array}{r} 825 \\ -160 \\ \hline \end{array}$$

**J**
$$\begin{array}{r} 3046 \\ -1506 \\ \hline \end{array}$$

## 2

**A**
$$\begin{array}{r} 492 \\ -286 \\ \hline \end{array}$$

**B**
$$\begin{array}{r} 903 \\ -791 \\ \hline \end{array}$$

**C**
$$\begin{array}{r} 681 \\ -491 \\ \hline \end{array}$$

**D**
$$\begin{array}{r} 7043 \\ -5110 \\ \hline \end{array}$$

**E**
$$\begin{array}{r} 76 \\ -38 \\ \hline \end{array}$$

**F**
$$\begin{array}{r} 3500 \\ -1350 \\ \hline \end{array}$$

**G**
$$\begin{array}{r} 9601 \\ -\ \ 490 \\ \hline \end{array}$$

**H**
$$\begin{array}{r} 6824 \\ -1904 \\ \hline \end{array}$$

**I**
$$\begin{array}{r} 584 \\ -466 \\ \hline \end{array}$$

**J**
$$\begin{array}{r} 745 \\ -\ \ 80 \\ \hline \end{array}$$

# Mastery Test Review—Lesson 24

## 1

**A**
$$12\begin{cases} 4 & \text{\_\_\_\_\_\_\_\_\_\_\_\_} \\ \square & \text{\_\_\_\_\_\_\_\_\_\_\_\_} \end{cases}$$

**B**
$$12\begin{cases} 2 & \text{\_\_\_\_\_\_\_\_\_\_\_\_} \\ \square & \text{\_\_\_\_\_\_\_\_\_\_\_\_} \end{cases}$$

**C**
$$12\begin{cases} 3 & \text{\_\_\_\_\_\_\_\_\_\_\_\_} \\ \square & \text{\_\_\_\_\_\_\_\_\_\_\_\_} \end{cases}$$

**D**
$$12\begin{cases} 5 & \text{\_\_\_\_\_\_\_\_\_\_\_\_} \\ \square & \text{\_\_\_\_\_\_\_\_\_\_\_\_} \end{cases}$$

## Mastery Test Review Lesson 24 (continued)

**2**

$$\begin{array}{r}12\\-\ 5\\\hline\end{array}\quad\begin{array}{r}12\\-\ 8\\\hline\end{array}\quad\begin{array}{r}12\\-\ 7\\\hline\end{array}\quad\begin{array}{r}12\\-\ 9\\\hline\end{array}\quad\begin{array}{r}12\\-\ 6\\\hline\end{array}\quad\begin{array}{r}12\\-\ 7\\\hline\end{array}\quad\begin{array}{r}12\\-\ 8\\\hline\end{array}\quad\begin{array}{r}12\\-\ 9\\\hline\end{array}$$

$$\begin{array}{r}12\\-\ 4\\\hline\end{array}\quad\begin{array}{r}12\\-\ 9\\\hline\end{array}\quad\begin{array}{r}12\\-\ 5\\\hline\end{array}\quad\begin{array}{r}12\\-\ 8\\\hline\end{array}\quad\begin{array}{r}12\\-\ 3\\\hline\end{array}\quad\begin{array}{r}12\\-\ 7\\\hline\end{array}\quad\begin{array}{r}12\\-\ 6\\\hline\end{array}\quad\begin{array}{r}12\\-10\\\hline\end{array}$$

## Mastery Test Review—Lesson 30

**1**

**A**

$12 \begin{cases} 5 \\ \square \end{cases}$ ----------------

**B**

$7 \begin{cases} 6 \\ \square \end{cases}$ ----------------

**C**

$\square \begin{cases} 9 \\ 1 \end{cases}$ ----------------

**D**

$\square \begin{cases} 7 \\ 3 \end{cases}$ ----------------

**E**

$7 \begin{cases} 2 \\ \square \end{cases}$ ----------------

**F**

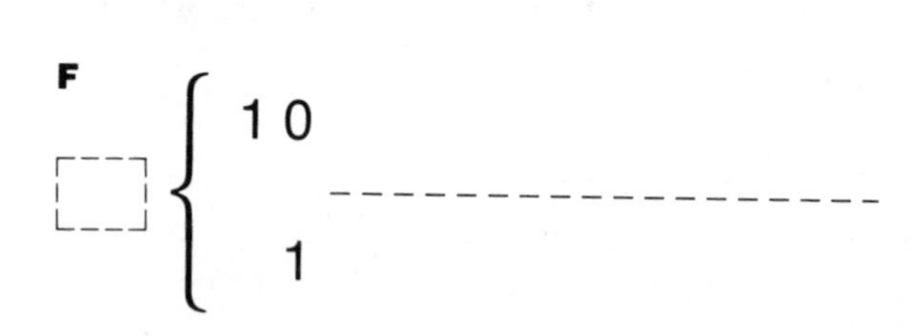

$\square \begin{cases} 10 \\ 1 \end{cases}$ ----------------

**2**

**A**

$9 \begin{cases} 8 \\ \square \end{cases}$ ----------------

**B**

$\square \begin{cases} 8 \\ 6 \end{cases}$ ----------------

Part 2 continues on the next page.

## Mastery Test Review Lesson 30 (continued)

C

☐ { 3, 9 } ----------

D

3 { 1, ☐ } ----------

E

☐ { 8, 4 } ----------

F

7 { 5, ☐ } ----------

## Mastery Test Review—Lesson 32

**1**

| A | B | C | D | E | F |
|---|---|---|---|---|---|
| 4006 | 5027 | 7004 | 2902 | 4098 | 7007 |

**2**

| A | B | C | D | E | F |
|---|---|---|---|---|---|
| 5009 | 9035 | 8024 | 7406 | 9004 | 8201 |

## Mastery Test Review—Lesson 33

**1**

A $\begin{array}{r} \underline{40}46 \\ -2185 \\ \hline \end{array}$

B $\begin{array}{r} 5024 \\ -1853 \\ \hline \end{array}$

C $\begin{array}{r} 3704 \\ -2198 \\ \hline \end{array}$

D $\begin{array}{r} 5029 \\ -3641 \\ \hline \end{array}$

E $\begin{array}{r} 904 \\ -186 \\ \hline \end{array}$

**2**

A $\begin{array}{r} 502 \\ -185 \\ \hline \end{array}$

B $\begin{array}{r} 308 \\ -157 \\ \hline \end{array}$

C $\begin{array}{r} 5049 \\ -\ \ 768 \\ \hline \end{array}$

D $\begin{array}{r} 7082 \\ -3140 \\ \hline \end{array}$

E $\begin{array}{r} 3068 \\ -\ \ 184 \\ \hline \end{array}$

# Mastery Test Review—Lesson 37

## 1

$$\begin{array}{r}16\\-\ 9\\\hline\end{array}\quad\begin{array}{r}11\\-\ 9\\\hline\end{array}\quad\begin{array}{r}14\\-\ 9\\\hline\end{array}\quad\begin{array}{r}12\\-\ 9\\\hline\end{array}\quad\begin{array}{r}16\\-\ 9\\\hline\end{array}\quad\begin{array}{r}13\\-\ 9\\\hline\end{array}\quad\begin{array}{r}17\\-\ 9\\\hline\end{array}\quad\begin{array}{r}15\\-\ 9\\\hline\end{array}$$

## 2

$$\begin{array}{r}10\\-\ 7\\\hline\end{array}\quad\begin{array}{r}10\\-\ 6\\\hline\end{array}\quad\begin{array}{r}10\\-\ 9\\\hline\end{array}\quad\begin{array}{r}10\\-\ 8\\\hline\end{array}\quad\begin{array}{r}10\\-\ 6\\\hline\end{array}\quad\begin{array}{r}10\\-\ 9\\\hline\end{array}\quad\begin{array}{r}10\\-\ 7\\\hline\end{array}\quad\begin{array}{r}10\\-\ 8\\\hline\end{array}$$

$$\begin{array}{r}10\\-\ 3\\\hline\end{array}\quad\begin{array}{r}10\\-\ 7\\\hline\end{array}\quad\begin{array}{r}10\\-\ 2\\\hline\end{array}\quad\begin{array}{r}10\\-\ 4\\\hline\end{array}\quad\begin{array}{r}10\\-\ 8\\\hline\end{array}\quad\begin{array}{r}10\\-\ 6\\\hline\end{array}\quad\begin{array}{r}10\\-\ 9\\\hline\end{array}\quad\begin{array}{r}10\\-\ 1\\\hline\end{array}$$

# Mastery Test Review—Lesson 43

## 1

**A** Our school had a dog show for German shepherds and hunting dogs. There were 15 German shepherds in the show. 19 dogs had been brought to the show. How many hunting dogs were there?

☐ ________________

**B** In the morning 16 jet planes left New York. In the afternoon 74 small planes left New York. How many planes in all left New York?

☐ ________________

**C** There are 7 rosebushes in our garden. We have 4 red rosebushes. The rest are yellow. How many yellow rosebushes do we have?

☐ ________________

Part 1 continues on the next page.

**D** There are 14 clean cups on the shelf. There are 19 dirty cups in the sink. How many cups are there in all?

☐ ____________________

**E** 14 women were in the play. 29 people were in the play. How many men were in the play?

☐ ____________________

**F** My aunt has 28 white flowers. She has 39 flowers in all. The rest of the flowers are pink. How many pink flowers does my aunt have?

☐ ____________________

## 2

**A** Bill painted 43 houses. He painted 52 stores. How many buildings did Bill paint?

☐ ____________________

**B** In the waiting room of a doctor's office there is a huge fish tank. There are 14 goldfish in the tank and the rest are sunfish. There are 30 fish in all in the tank. How many sunfish are in the tank?

☐ ____________________

**C** There are 40 girls in the kindergarten of a school. There are 75 children in kindergarten. How many are boys?

☐ ____________________

**D** Miss Manos is selling tickets. She sold 31 circus tickets this morning. This afternoon she sold 50 movie tickets. How many tickets in all did she sell?

☐ ____________________

Part 2 continues on the next page.

## Mastery Test Review Lesson 43 (continued)

**E** Marcia Kabatie has entered many skating contests. 43 were ice-skating contests and the rest were roller-skating ones. She has taken part in 60 contests in all. How many roller-skating contests has she entered?

[ ] ____________________

## Mastery Test Review—Lesson 44

**1**

| 9 | 9 | 9 | 9 | 9 | 9 | 9 | 9 |
|---|---|---|---|---|---|---|---|
| − 7 | − 4 | − 3 | − 6 | − 2 | − 8 | − 7 | − 1 |

| 9 | 9 | 9 | 9 | 9 | 9 | 9 | 9 |
|---|---|---|---|---|---|---|---|
| − 5 | − 6 | − 4 | − 7 | − 3 | − 6 | − 2 | − 7 |

**2**

| 13 | 13 | 13 | 13 | 13 | 13 | 13 | 13 |
|---|---|---|---|---|---|---|---|
| − 7 | − 9 | − 4 | − 8 | − 5 | − 6 | − 4 | − 7 |

| 13 | 13 | 13 | 13 | 13 | 13 | 13 | 13 |
|---|---|---|---|---|---|---|---|
| − 6 | − 8 | − 4 | − 5 | − 9 | − 7 | − 6 | − 8 |

# Mastery Test Review—Lesson 48

**A** There are 143 cabins up at the lake. This year more cabins were built. Now there are 160. How many more cabins were built at the lake?

**B** My dad's store was having a sale. There were 114 people in the store. Some more people came in. Then there were 142 people in the store. How many more people came into the store?

**C** Benji had a stamp collection. When the year began, he had 1436 stamps. During the year he got 148 stamps. How many stamps did Benji have at the end of the year?

**D** A meal was served at a banquet. There were 143 people sitting at the tables. Some more people arrived. Now there are 160 people at the tables. How many more people arrived?

**E** When the game started, there were 842 people watching it. During the game another 58 people began to watch it. How many people were watching the game altogether?

## 2

**A** When the movie began, there were 146 people in the theater. By the end of the movie there were 206 people in the theater. How many people came into the theater after the movie started?

[ ] ________________

**B** Last summer we took photographs of a building that was being built. At that time the building had 58 floors. In winter the building was finished. It had 70 floors. How many floors were built after we took photographs?

[ ] ________________

**C** 315 children took the bus to summer camp. Another 46 children took the train to camp. How many children went to summer camp in all?

[ ] ________________

**D** At a beach there were 54 swimmers. 40 more swimmers came. How many swimmers were at the beach then?

[ ] ________________

**E** A dress shop had 23 dresses. Some more dresses came in a truck. Now there are 70 dresses in the shop. How many dresses came in the truck?

[ ] ________________

# Mastery Test Review—Lesson 50

**1**

| | | | | | | | |
|---|---|---|---|---|---|---|---|
| 7 − 2 | 8 − 2 | 6 − 2 | 10 − 2 | 8 − 2 | 5 − 2 | 11 − 2 | 9 − 2 |
| 5 − 2 | 10 − 2 | 6 − 2 | 11 − 2 | 7 − 2 | 8 − 2 | 4 − 2 | 9 − 2 |

**2**

| | | | | | | | |
|---|---|---|---|---|---|---|---|
| 15 − 6 | 15 − 10 | 15 − 7 | 15 − 9 | 15 − 8 | 15 − 6 | 15 − 9 | 15 − 7 |
| 15 − 10 | 15 − 7 | 15 − 6 | 15 − 9 | 15 − 8 | 15 − 7 | 15 − 6 | 15 − 9 |

# Mastery Test Review—Lesson 53

**1**

**A** 5000 − 3

**B** 5000 − 30

**C** 5000 − 300

**D** 8000 − 4

**E** 8000 − 40

**F** 3000 − 8

**G** 3000 − 800

**H** 1000 − 7

**I** 1000 − 20

**J** 1000 − 400

## Mastery Test Review Lesson 53 (continued)

**2**

A
$$\begin{array}{r} 3000 \\ -\quad 8 \\ \hline \end{array}$$

B
$$\begin{array}{r} 3000 \\ -\ 800 \\ \hline \end{array}$$

C
$$\begin{array}{r} 3000 \\ -\quad 80 \\ \hline \end{array}$$

D
$$\begin{array}{r} 1000 \\ -\quad 20 \\ \hline \end{array}$$

E
$$\begin{array}{r} 1000 \\ -\quad 2 \\ \hline \end{array}$$

F
$$\begin{array}{r} 4000 \\ -\ 245 \\ \hline \end{array}$$

G
$$\begin{array}{r} 4000 \\ -2450 \\ \hline \end{array}$$

H
$$\begin{array}{r} 8000 \\ -\ 342 \\ \hline \end{array}$$

I
$$\begin{array}{r} 8000 \\ -\ 300 \\ \hline \end{array}$$

J
$$\begin{array}{r} 6000 \\ -4128 \\ \hline \end{array}$$

## Mastery Test Review—Lesson 54

**1**

**A** 90 children went to a play. 46 of the children enjoyed the play. The rest did not. How many of the children did not enjoy the play?

**B** We bought some eggs. When we got home, 36 eggs were broken. 24 eggs were not broken. How many eggs in all did we buy?

**C** There are 62 sharks. 45 of the sharks are hungry. How many of the sharks are not hungry?

**D** 80 people were watching a bulldozer knock down a building. 69 people left to return to work. How many people stayed to watch?

Part 1 continues on the next page.

**E** Linda Arrowhead's father is a baker. One Saturday Linda helped her father make pies for some restaurants. They made 60 apple pies and 45 pies that were not apple. How many pies in all did Linda and her father make?

**A** My sister made a beaded belt. She used 140 large beads to make the belt. 90 of the beads are red. How many of the beads are not red?

**B** There's a new statue in our city. Lots of people saw the statue. 80 people liked the statue. 57 people did not like the statue. How many people saw the statue?

**C** There are 140 people on the beach. 95 of them are wearing hats. How many of the people are not wearing hats?

**D** There were 86 earthquakes last year. 29 of the earthquakes were in Europe. How many earthquakes were not in Europe?

**E** We made sandwiches for our club's picnic. The people who came ate 95 ham sandwiches and 120 chicken sandwiches. How many sandwiches were eaten at the picnic?

# Mastery Test Review—Lesson 58

**1**

**A**

11 { 4 ______ / [ ] ______

**B**

11 { 2 ______ / [ ] ______

**C**

11 { 5 ______ / [ ] ______

**D**

11 { 3 ______ / [ ] ______

**2**

| 11 | 11 | 11 | 11 | 11 | 11 | 11 | 11 |
|---|---|---|---|---|---|---|---|
| − 8 | − 2 | − 7 | − 4 | − 6 | − 9 | − 3 | − 7 |

| 11 | 11 | 11 | 11 | 11 | 11 | 11 | 11 |
|---|---|---|---|---|---|---|---|
| − 6 | − 4 | − 8 | − 6 | − 5 | − 7 | − 3 | − 8 |

# Mastery Test Review—Lesson 60

**1**

**A** Mrs. Savas is 42 years old. Miss Hark is 70 years old. How many years older is Miss Hark?

**B** After a few games of table tennis, Amos Silverheels had scored 26 points. His brother Mike had scored 40 points. How many points did Amos and Mike score in all?

**C** Gina weighs 48 kilograms. Sam weighs 29 kilograms. How many more kilograms does Gina weigh?

Part 1 continues on the next page.

**D** Mel exercised for 48 minutes. Brad exercised for 42 minutes. How many minutes did Mel and Brad exercise in all?

**E** A building is 36 meters tall. A tree near the building is 48 meters tall. How many meters taller is the tree?

**A** Jana and Allison like to play marbles. Jana has 137 marbles. Allison has 170 marbles. How many more marbles does Allison have than Jana?

**B** My mother's new car weighs 1340 kilograms. My mother's old car weighed 1700 kilograms. How many kilograms do the cars weigh in all?

**C** Mr. Mills is 45 years old. Mr. Young is 60 years old. How many years older is Mr. Young?

**D** Rafael and Matt worked as ticket sellers at a circus. Rafael sold 94 tickets and Matt sold 120 tickets. How many tickets did the boys sell?

**E** An apple tree is 594 centimeters tall. An oak tree is 730 centimeters tall. How many centimeters taller than the apple tree is the oak tree?

# Mastery Test Review—Lesson 63

## 1

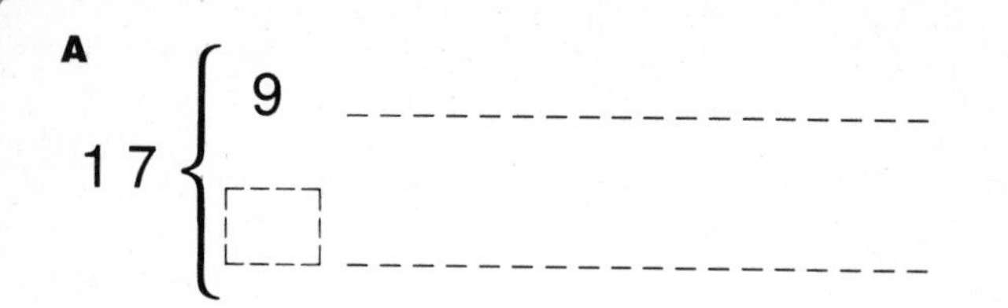

**A** 17 { 9 ________ / [ ] ________

**B** 14 { 9 ________ / [ ] ________

**C** 15 { 9 ________ / [ ] ________

**D** 16 { 9 ________ / [ ] ________

## 2

$15 - 9 =$ ____ $\quad 17 - 8 =$ ____ $\quad 12 - 6 =$ ____ $\quad 10 - 5 =$ ____ $\quad 16 - 7 =$ ____ $\quad 14 - 7 =$ ____ $\quad 14 - 9 =$ ____ $\quad 15 - 6 =$ ____

$15 - 7 =$ ____ $\quad 14 - 5 =$ ____ $\quad 14 - 6 =$ ____ $\quad 14 - 7 =$ ____ $\quad 17 - 8 =$ ____ $\quad 17 - 9 =$ ____ $\quad 16 - 7 =$ ____ $\quad 14 - 5 =$ ____

# Mastery Test Review—Lesson 64

## 1

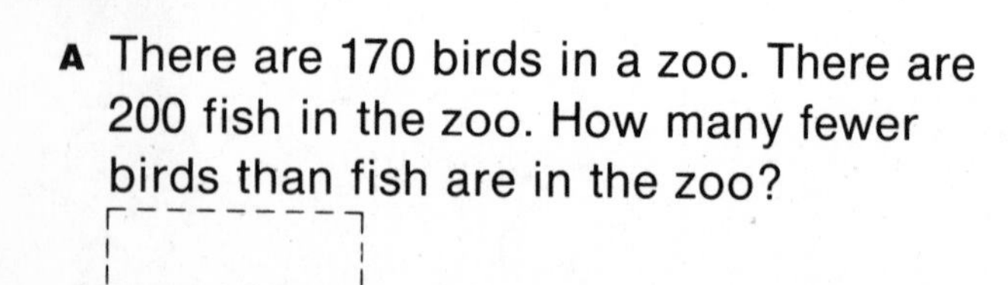

**A** There are 170 birds in a zoo. There are 200 fish in the zoo. How many fewer birds than fish are in the zoo?

[ ] ________

**B** We saw lots of squirrels and rabbits. We counted 143 squirrels and 180 rabbits. How many animals did we see?

[ ] ________

Part 1 continues on the next page.

**C** The cook let us choose cake or ice cream for dessert. 145 kids ate cake and 200 kids ate ice cream. How many more kids ate ice cream than cake?

[ ] ________________

**D** Iris taught 36 children how to jump rope. Christina taught 40 children how to jump rope. How many children learned how to jump rope?

[ ] ________________

**E** Chris has 47 rocks. Judd has 82 rocks. How many more rocks does Judd have than Chris?

[ ] ________________

**F** There are 90 bottles of soda pop. 72 of the bottles are full. How many bottles are not full?

[ ] ________________

**G** 1350 railway cars are full. 1500 railway cars are empty. How many railway cars are there in all?

[ ] ________________

**H** The Fireside Bookshop sold 1826 books in December. The shop sold 1900 books in January. How many books did the shop sell?

[ ] ________________

**I** A hockey team lost 42 games. It played 70 games in all. How many games did the hockey team win?

[ ] ________________

Part 1 continues on the next page.

**J** Fran ran 37 kilometers this week. Carol ran 40 kilometers this week. How many kilometers did Fran and Carol run this week?

[ ] ____________________

**K** Louise made 49 small cakes with strawberry frosting. Then she made some cakes with chocolate frosting. She baked 72 cakes in all. How many cakes had chocolate frosting?

[ ] ____________________

**L** Ann read 167 pages of a book. On her day off she read some more. Now Ann has read 210 pages. How many pages did Ann read on her day off?

[ ] ____________________

## 2

**A** 64 people walked around the ladder. 140 people walked under the ladder. How many more people walked under the ladder than around the ladder?

[ ] ____________________

**B** In one week 148 new cars and 473 old cars were washed at the Happy Car Wash. How many cars in all were washed?

[ ] ____________________

**C** At a camp 3802 oranges were eaten in one week. 1648 apples were also eaten. How many more oranges than apples were eaten?

[ ] ____________________

Part 2 continues on the next page.

**D** Andy put 2509 boxes on trucks. Randy put 3108 boxes on trucks. How many fewer boxes did Andy put on trucks than Randy?

[ ] ________________

**E** Workers chopped down 185 oak trees and 190 pine trees. How many trees were chopped down?

[ ] ________________

**F** Last spring a shop had a sale on kites. The shop sold 158 red kites and 190 yellow kites. How many kites did the shop sell?

[ ] ________________

**G** Lauren planted 97 flowers. Then she planted some more flowers. Now 102 flowers are planted. How many more flowers did she plant?

[ ] ________________

**H** A newspaper stand received 1444 magazines. 98 of the magazines had torn pages. How many magazines did not have torn pages?

[ ] ________________

**I** 148 people were ice-skating. Later, more people came to skate. Now there are 190 skaters. How many people came late to skate?

[ ] ________________

Part 2 continues on the next page.

**J** Phil danced for 148 minutes. Then he danced for another 52 minutes. How many minutes did he dance in all?

[ ] ____________________

**K** Workers cleaned 18 airplanes. They cleaned 24 boats. How many things did the workers clean?

[ ] ____________________

**L** An office had 1405 packages of paper. The office used up 1289 packages. How many packages of paper were left?

[ ] ____________________